MW01093535

JAGDTIGER

JAGDTIGER

The Most Powerful Armoured Fighting Vehicle of World War II
Operational History

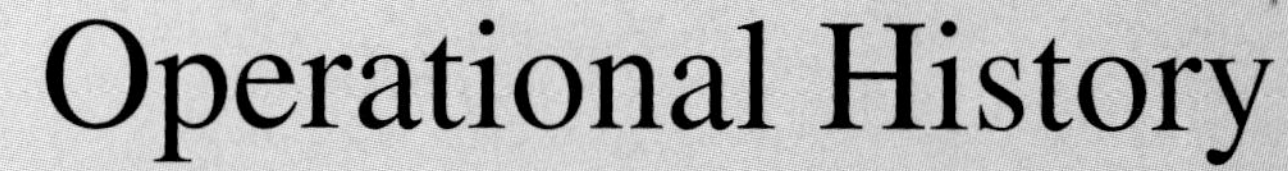

Andrew Devey

Schiffer Military History
Atglen, PA

ACKNOWLEDGEMENTS

The author wishes to express his thanks to the following contributors:

The staff at the Tank Museum especially David Fletcher, for providing the bulk of the technical photographs and mechanical information.

Walter J. Spielberger for checking through the technical analysis and permission for the use of his photographs.

Dr. Heinz Maus and Peter Gudgin for their technical help and Paul Cox for his technical edit of section one.

Wolfgang Schneider for his photographs, tactical information and veteran contacts.

Jean Paul Pallud for his assistance with the issues concerning the Ardennes offensive.

Karlheinz Münch for his help with the very detailed research into the schwere Panzer Jäger Abteilung 653, and the use of his exceptional photographs.

Tom Jentz and Werner Liebknecht for assistance in locating the documentation from the General Inspector of the Panzer Troops, 1944-1945, which are in the Militararchiv, Freiburg.

Kenneth H. Powers, U.S. Army, for help with U.S. intelligence information and Bill Auerback who helped with U.S. Archives Photographs.

Thomas Anderson for providing Soviet information. Per Sonnervik for arranging our visit to Kubinka, in 1993.

Heiner F. Duske for help with veteran contacts. The two former commanders, Walter Scherf and Otto Carius, for their assistance in clarifying the anomalies concerning the schwere Panzer Jäger Abteilung 512.

Erika and Harold Scrivens and Janie Chapman for help with translations, Richard D Ord and my father Ronald Devey for proof reading.

My wife Joanne and daughter Heather for their support and to whom I dedicate this book.

PHOTO CREDITS

Archives
The Tank Museum, Bovington Camp, Wareham, Dorset, BH20 6JG
The Bundesarchiv (BA) Postfach 320, 5400 Koblenz 1, Federal Republic of Germany
The National Archives (U.S. Army) National Archives Trust Fund Board, Washington DC 20408
Establissement Cinematographigue et Photographigue Des Armees, (ECPA) Fort D'Ilvry, 75998 Paris, France
The Bundesarchiv Militararchiv (Militararchiv), Wiesental Strasse, 10, 7800 Freiburg, Germany
The Delta Publication Co., Ltd. (Delta Publishing), 1-50 Kanda Jinbocho, Chiyoda-ku, Tokyo 101, Japan
Bastogne Historical Research Center

Individuals
Thomas Anderson
Bill Auerbach
Peter Gudgin
Karlheinz Münch
Jean Paul Pallud
Wolfgang Schneider
Per Sonnervik
Walter J. Spielberger

Drawing credits
The Tank Museum
Bundesarchiv Military Archiv Freiburg
Author

Documents
Bundesarchiv Military Archiv Freiburg
Karlheinz Münch

Maps and color artwork
Author

Plate 1. (*pages 2-3*) The last three Jagdtigers and the fighting troops of 1/s.Pz.Jg.Abt 512 in the market square in Iserlohn on 16.4.1945 (U.S. National Archives).

Book Design by Ian Robertson.

Library of Congress Catalog Number: 98-88850

Printed in China.
ISBN: 0-7643-0751-7

We are interested in hearing from authors with book ideas on related topics.

Published by Schiffer Publishing Ltd.
4880 Lower Valley Road
Atglen, PA 19310
Phone: (610) 593-1777
FAX: (610) 593-2002
E-mail: Schifferbk@aol.com.
Visit our web site at: www.schifferbooks.com
Please write for a free catalog.
This book may be purchased from the publisher.
Please include $3.95 postage.
Try your bookstore first.

In Europe, Schiffer books are distributed by:
Bushwood Books
6 Marksbury Road
Kew Gardens
Surrey TW9 4JF
England
Phone: 44 (0)181 392-8585
FAX: 44 (0)181 392-9876
E-mail: Bushwd@aol.com.

Try your bookstore first.

Contents

22

European War Situation 1944-1945

Before detailing the experiences of the Combat Units with their Jagdtigers, we must first look at the overall situation facing the German Army in this final year of World War II.

22.1 Eastern Front

With the start of the Soviet summer offensive in Byelorussia in June 1944, the German High Command were fooled, by an elaborate Soviet deception, into believing that the main attack would come from the Ukraine salient, and attack Germany's Romanian and Bulgarian Allies.

The Soviet offensive hit a depleted Army Group (Center) in June 1944, with 5,200 tanks and assault guns. This attack was able to surround and destroy Army Group (Center) which then opened the way into Poland and East Prussia.

Through this breach in the Front, the Soviets launched repeated attacks against the flanks, pushing deep into Romania and the Western Ukraine. In the late summer and autumn further, Soviet attacks cleared most of the Eastern Baltic coast. The Soviet offensive ran out of steam because it had suffered very high casualties along with shortages of ammunition and fuel supplies.(1)

Fighting continued all along the Front throughout the autumn of 1944. The Soviets managed to conserve and indeed create a new armored force, which was ready for the final attacks to start in January 1945.

The Soviet main thrust towards Berlin involved over 7,000 tanks and assault guns; this compared to Germany's 1,200 tanks and assault guns. As well as having a numerical advantage, the Soviet tank crews were also matching the training and skills of their German opponents.

With the loss of the Romanian oil fields, in autumn 1944, the Germans suffered extreme fuel shortages as well as the lack of all other supplies. Vicious air attacks by the Allied and Soviet air forces, directed at the road and rail infrastructure disrupted the German's tactical movements.

In mid-January 1945, the Soviets attacked again in the North, with their East Prussia Offensive and in the center with the Vistula-Oder Offensive.

At this stage of the War, the German Army's morale was very low. Hitler needed a propaganda victory, to boost the morale of his battered forces, and launched the ill-fated attack at Lake Balaton. The offensive failed after two weeks of bitter fighting in treacherous weather.

After this abortive German offensive, repeated Soviet offensives, notably the Vienna Offensive followed by the Berlin and Prague Offensives, pushed the German forces backwards.

The Red Army had the last firefights against Jagdtigers, in Austria, in early May 1945, the only time Jagdtigers fought against Soviet forces.

22.2 Western Front

The long awaited Allied attack to establish a Second Front in Northern Europe started at 0630 hours on 6 June 1944, with the Allied landings on the beaches of Normandy. The Germans thought that this was a diversionary tactic with the main landings to take place near Calais, hence the bulk of German armor was kept in reserve to counter this expected main attack. Bitter fighting continued throughout June to establish a more extensive bridgehead.

In mid-July, the British launched "Operation Goodwood" to try to push south of Caen. This pinned down the heavily reinforced Panzer units. By late July, the Allies had nearly two million troops and 150,000 vehicles in France.

The Allied breakout from Normandy occurred when Patton's third Army over-ran Brittany. Allied troops forced the German armor into a pocket between Falaise and Argentan. The destruction of Army Group B was almost complete in the pocket with 344 German tanks and assault guns, 2,500 soft-skinned vehicles and 252 guns either destroyed or abandoned. With the catastrophic loss of 1,500 tanks and assault guns in the Normandy fighting, the Germans could not contain the sweeping Allied advances, which liberated huge parts of France and Belgium.

In September 1944, the British, with the aid of paratroop landings, tried to capture a bridge over the Rhine at Arnhem. They captured a bridge at Nijmegen, but the attack failed in its ultimate objective. By 15 December, the Allies were ready to cross of the Rhine but German resistance was strong.

In mid-December, the Americans settled down in the Ardennes, expecting the slow advance into Germany to continue. However, Hitler had other plans. He launched a major offensive, "Wacht am Rhein", that was to break through the Ardennes, separate the British and American forces and capture Antwerp. With almost 1,000 tanks and assault guns, it nearly succeeded, but with empty fuel-tanks and increasing opposition from Allied troops, the offensive ground to a halt.

After the failure of this offensive, while the Allies attacked the bulge created, a second German offensive launched on 1 January 1945, in the Alsace area, to try to recapture Strasbourg. It was this second offensive that saw the use of Jagdtiger in combat (Platoon strength) for the first time.

Through February to 25 March, Allied attacks along the whole Front from Strasbourg to Arnhem, ended the German force on the West Bank of the Rhine. With the capture of the Rail Bridge at Remagen, on 7 March, the Americans formed a huge bridgehead east of the river.

The continuous Allied advances created two large pockets of German troops and equipment in the Ruhr area and the Harz mountains.

During April, the Allies might have reached Berlin itself but Eisenhower kept to his agreement with the Soviets that they would halt at the Elbe. In April and May Patton's third Army swept south and east into Austria to link with the Soviets near Linz on 5 May 1945.

22.3 Italy

At the end of 1943, the Allied forces in Italy were in a stalemate situation and early hope of being in Rome for Christmas had faded in November.

In January 1944, the Americans launched an amphibious assault at Anzio to outflank the Germans and open the coast road. The attack involved 50,000 British and American troops under Major General Lucas. The landing was virtually unopposed but an over-cautious Lucas gave the German commander, Kesselring, time to reinforce the Anzio perimeter with 8 German divisions.

The Allies were having no more luck against the main offensive line. Three attacks against the defenders of Monte Cassino were fruitless.

An attempt in February to breakout of Anzio also failed and German counter attacks almost pushed the Allies back into the sea. This stalemate continued for two months through heavy rains.

The 11 May, saw a renewed attack at Cassino, which succeeded. At the same time, the Allies launched another attack at Anzio. This strike opened the way to Rome, which Kesselring declared an open city on 4 June, to prevent its destruction. The German forces withdrew 150 miles to the Gothic Line.

Because of events elsewhere, some of General Alexander's troops withdrew from Italy and took part in the invasion of Southern France, "Operation Dragoon". This removed the impetus from the future offensives in the northern thrust towards the Alps.

The offensive against the Gothic Line started in September and broke through at three points. Kesselring managed to stem the breaches, achieving a stalemate until April 1945.

The final offensive opened on 9 April 1945, when the British eighth Army crossed the River Senio, followed shortly by the American fifth Army capturing Bologna. At this time, the Soviets were fighting in Berlin and the Allies were capturing huge areas of Western Germany. The surviving German troops in Italy quickly lost their morale.

American troops entered Milan on the 29 April. Following this, the German troops in Italy surrendered unconditionally. No Jagdtigers saw action on this front.

(1)It is noteworthy that the Soviets lost almost 25,000 tanks and assault guns during fighting in 1944. Coincidentally in that same year the Soviets started to concentrate production facilities on the heavier IS-2 and larger assault guns ISU-122 and ISU-152 to counter the German Panthers and Tigers and their variants. The IS-2 was roughly comparable to the German Panther tank.

Generally the German tanks had superior technology, in guns and ammunition compared to the same caliber weapons of their opponents. The Soviet answer to this was to put a bigger gun on a light chassis.

23

Jagdtigers for Testing and Training

23.1 Factory testing

Before the release of a vehicle from the factory, it underwent a series of quality inspections, by the factory's own Quality Control Department. These checks included examination, measurement and test-driving, but no gunnery trials.

The first running trials of the Jagdtiger chassis, was at Nibelungen Werk with the mild-steel prototype Porsche vehicle (Chapter 15, Plate 123 *see Jagdtiger Vol. I*), testing of the Porsche suspension system being achieved?

Completion of the first two Jagdtigers occurred in February 1944. After initial testing in the factory grounds, they were sent by rail to Arys Proving Ground, to be demonstrated for the Fuhrer, on 20 April 1944, his birthday. Only one Jagdtiger ran on the demonstration.

The two Jagdtigers, 305001 (Porsche) and 305002 (Henschel) then traveled by rail, to Kummersdorf for independent evaluation by the Army weapons testing department, Waffen Pruef 6. They arrived during May 1944, and were both photographed at the test facility.

23.2 Independent testing

As with any new vehicle, whether commercial or military in design, after initial production, the first few vehicles require testing to prove the design.

During World War II, the German Army's vehicle testing operation was Waffen Pruef 6 (Weapons Testing 6). This test establishment was at Kummersdorf, an area within the huge tank training area of Sennelager, situated to the North of Paderborn. One of its most recognizable features, was the large concrete wading pool at Haustenbeck.

When a vehicle transferred to the test center, it was assigned an identification number which was paint stenciled onto its front glacis plate. All reports refer to the test number and not the vehicle's chassis number.

Plate 178. Side view of Porsche Jagdtiger chassis number 305001 in Kummersdorf undergoing evaluation in May 1944. No provision for external stowage had been fitted to the vehicle sides at this stage (Tank Museum).

Plate 179. Taken at virtually the same time, the Henschel Jagdtiger chassis number 305002 during the same trials for evaluation by Waffen Pruef 6 (BA).

The Jagdtigers tested by Waffen Pruef 6 had the following numbers:

Chassis No:	Wa. Pru. 6 No:
305001	221
305002	222
305004	253

Trials on the first two Jagdtigers lasted for a six-week period. Ing. Escher, the test engineer in charge of this project, sent a report, dated 21 July 1944, stating that the trials had been inconclusive after the six week period and that further testing was necessary.

The reason for the unfinished test series was due to Jagdtiger (No 305001). It sustained damage during testing and could not be repaired at Kummersdorf. It required to be returned to Nibelungen Werk for repair. Therefore, a second Porsche Jagdtiger was delivered to Kummersdorf. Jagdtiger No. 305004, completed in July, was chosen. The only other Jagdtiger available at the time was No. 305005; this had failed a quality inspection at the end of July, and was therefore not acceptable.

Jagdtiger (No. 305004) was delivered to Kummersdorf in August 1944, for testing to be concluded. Interestingly, this is not on the Nibelungen release figures. The Nibelungen Werk record, amended on 1 September 1944, records this. Jagdtiger (No 305004) was to remain in the Kummersdorf area and used extensively for trials, including towing, throughout the winter of 1944/45.

The trials favored the Henschel suspension system, which was more robust and the decision was taken to stop the Porsche suspension production, and continue on the Henschel system vehicles. The last Porsche vehicle (No 305012), was completed in September 1944.

Plate 180. Front view of Jagdtiger chassis number 305004 undergoing testing in Kummersdorf in September 1944 (Walter J. Spielberger).

23.3 The first training vehicle (Jagdtiger No 305003)

The third Jagdtiger, completed in June 1944, was used at Nibelungen Werk, to evaluate the feasibility of interchanging track parts between the Ferdinand and Jagdtigers. To this end, there was a set of modified drive-sprockets drilled to accept 21 bolts, to fit the special 21 teeth drive-sprockets. These were required to engage the smaller pitch, and narrower, Kgs. 64/140/130 Ferdinand tracks. The other factor was that Porsche was not fully satisfied with flame cutting to remove the inner guide-horn from the Henschel type tracks. These had given track guide problems with the Henschel rear-idler. The flame cutting could soften the track links.

The trials proved the need to continue with the Henschel tracks, for greater combat reliability. These were more robust and gave the whole vehicle a 20% lower ground pressure, when comparing the two. This decision resulted in the modification of the return-idler. This gave better track-captivation with Ferdinand and modified Henschel Jagdtiger tracks, to suit the Porsche suspension vehicles.

Plate 181. The test engineers put No 305004 through its paces at the proving grounds. Right-side view shown (Walter J. Spielberger).

Plate 182. Left-side view, the vehicle still has its full suspension system (Walter J. Spielberger).

Plate 183-184. Two further views of Jagdtiger chassis number 305004 (Walter J. Spielberger).

After these running trials, transport tracks were fitted to Jagdtiger (No 305003). It was dispatched by rail on 30 June 1944, to the training area at Mielau, where a Panzer Jäger course was planned to train future Jagdtiger crews. Because of these trials it was not recorded as being completed, until July 1944, in the Nibelungen Werk records.

This was the first Jagdtiger with an application of Zimmerit ant-magnetic mine coating. It had a distinctive ring around its ball mount and was the first Jagdtiger to be fitted with hooks to carry spare track links. It had no external gun support.

23.4 The second training vehicle (Jagdtiger No. 305005)

The fifth Jagdtiger also a Porsche-suspension vehicle, was completed in Nibelungen Werk in July 1944.

After a thorough examination at the factory, an internal letter was circulated stating that there was a problem with the armor composition of its superstructure front. Therefore, the decision was taken, on 31 July 1944, to reject the vehicle. It was not to be issued to combat troops.

The Jagdtiger remained at Nibelungen Werk until 14 October 1944, it was then transported North to Putlos, the gunnery school on the Baltic coast, and this was an outstation of the Training and Reserve Battalion 500.

Jagdtigers 305006 onwards, were issued direct to the Ersatzheer (Army Reserve) and then to the combat units!

Other Jagdtigers were issued or transferred to different training centers, these are covered in Chapter 40.

Plate 189-190. Two views of Jagdtiger chassis number 305003 being run on transport tracks, in late June 1944, prior to being sent on to Mielau. This vehicle had a distinctive ring around the bow machine-gun dome (Karlheinz Münch).

Plate 185-188. Four views of Porsche Jagdtiger chassis number 305003 being tested at Nibelungen Werk. It is running on Ferdinand tracks. The vehicle required modified drive sprockets and teeth rings. It is also testing the modified rear-idler configuration. Photographs were taken in early June, 1944 (Tank Museum).

24

Jagdtiger Deployment Analysis

24.1 Jagdtiger deployment

By the time Jagdtigers were being steadily delivered for action, the war situation for Germany was hopeless. Hitler had depleted the Eastern Front of some of its armor, only to lose it in the Ardennes Offensive. The Luftwaffe had virtually been shot out of the skies, leaving the Allied and Soviet Air Forces in command of the airspace.

The once mighty Panzers were now impotent against the rockets and bombs from the air. Skilled camouflage and night operations were armor's only protection. The German Army's morale was very low, particularly on the Western Front (except for a few die-hard SS units).

With only 88 Jagdtigers built, their overall tactical importance was minimal. The majority were used in their intended role as movable strong points. The Jagdtigers' mobility in the field was restricted for several reasons, amongst which the most important was fuel shortage. Because of their extreme weight they were highly susceptible to mechanical breakdowns. They also presented a slow moving and relatively easy target to Allied fighter-bombers. Because of these restrictions, all long journeys were made by rail, with trains only traveling by night and remaining either in tunnels or in wooded areas during daylight hours.

The rail routes for all journeys had to be carefully considered to utilize only those bridges capable of carrying these heavy vehicles. Routes had to be continually reappraised because of the severe transport disruption caused by repeated air attacks on the German road and rail infrastructure.

In case of any vehicle suffering mechanical breakdown to the extent that retrieval was impossible, the crews had precise orders for the thorough destruction of the tank to prevent capture by the enemy. Charges specifically designed for this purpose were carried in the tank at all times and were placed under the engine and under the gun traversing mechanism. This effected a thorough self-destruction as later photographs illustrate.

Plate 191. Jagdtiger Chassis number 305007 prepared for rail transport has just been driven off a transport train in Fallingbostel shunting yard, on 9 September 1944 (Wolfgang Schneider).

24.2 Jagdtigers deployed

Field deployment of the 88 Jagdtigers was as follows:

Test vehicle	Wa. Pruef 6		Factory	(305001)	1	(1. Porsche)
Training vehicle	Wa. Pruef 6		Freistadt	(305002)	1	
Training vehicle	Mielau		Freistadt	(305003)	1	(1. Porsche)
Test Vehicle	Wa. Pruef 6			(305004)	1	(1. Porsche)
Training vehicle	Putlos Gunnery School			(305005/305049)	2	(1. Porsche)
s.Pz.Jg.Abt. 653						
Training vehicle	Ersatzheer	- 653 -	Ersatzheer		1	(1 Porsche)[1]
	3/653	Alsace	Germany		14	
	1/653	Alsace	Germany		14	
	2/653	Alsace	Germany		14	
	653 total				42[2]	
s.Pz.Jg.Abt.512						
	2/512	Ruhr			10	
	1/512	Ruhr			10	
	3/512	Sennelager/Harz			5	
	3/512	Rail transit -?			2[3]	
	512 total				27	
6 SS Panzer Armee						
	653/LAH	Linz			4	
Factory						
	653	Linz (not released)			8	(8.8 cm.)[4]
		Total			88	(11. Porsche)

Plate 192. Jagdtiger chassis number 305004 on a Gotha 80 ton trailer being taken to England for evaluation, in mid 1945 (Tank Museum).

24.3 Transport for Jagdtigers

By rail

All long journeys for Jagdtigers were undertaken by rail, with special tank transporting flatcars being used. The tank had to be prepared for carriage, by removing the skirt armor and swapping the battle tracks for transport tracks. This track change reduced the overall width from 3.67m to 3.27m. The skirt-armor and the battle tracks were loaded with the tanks.

To immobilize the tank on the special railway flat it was necessary to block the road wheels on both sides with wooden wedges; these placed between the tracks and the road wheels. This was sufficient to prevent the tank's movement even in the event of the flat's severe jolting during shunting.

No more that six Jagdtigers were permitted to be loaded on one train. These were interspersed with two flatcars to avoid overloading the trains, and presumably to avoid overloading bridges, etc.

Soft skinned vehicles such as lorries or reconnaissance vehicles were loaded onto other flats. In some circumstances, short journeys near to the front line, Jagdtigers were sent forward on rail flats still fitted with combat tracks. See Chapter 16.9 in volume I for changing tracks.

By road

Jagdtigers could be moved on roads when loaded onto "Gotha" 80 - metric tons tank transport trailers. With this method of transport, it was not practical to move over 10km/hour. This system was therefore only used for recovery purposes and not fast tactical movement.

There was no need to change tracks for this transport system, as the trailers were capable of taking a Jagdtiger at its full width.

Plate 193 - 195. Jagdtiger chassis number 305004 being used for towing tests in Kummersdorf, in January 1945. It still has a full set of suspension units and is seen here pulling a Tiger II (Walter J. Spielberger).

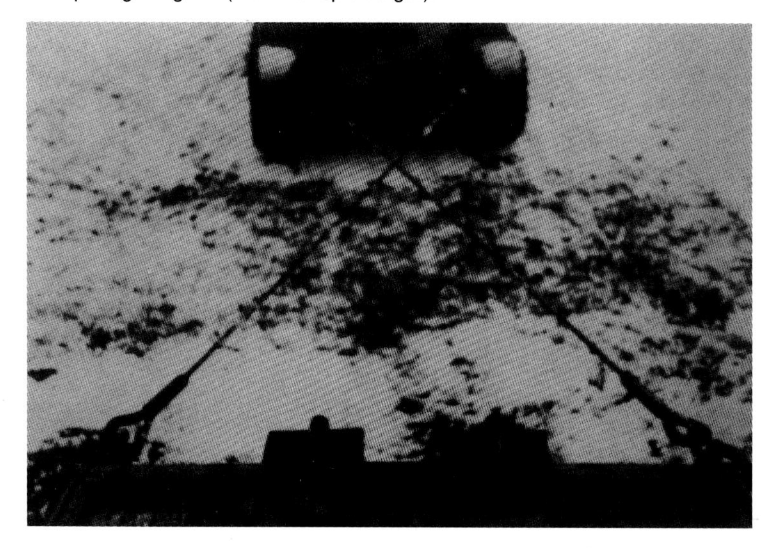

Towing

For recovery purposes Jagdtigers could be towed using a Berge Panther together with two 18 metric ton Half-tracks.

Directions for towing:

Instructions: When the steel cables were rolled or stretched they should not have been allowed to knot or kink. Steel cables were only to be connected using the correct hooks, shackles or eyes.

Directions for towing:

Before use, steel cable, was to be laid flat out on the ground.

The vehicle was not to be started with a jerk.

The cables were never bent around thin anchor pins or sharp edges.

After use, the cable it was cleaned and wound into a coil.

The cable was always kept oiled or greased to stop rusting.

Towing equipment stowed on vehicle:

4 'D' links;
2 'C' hooks;
2 steel cables 8.2m long at 32mm diameter.

Specific orders were issued to prevent the attempted recovery of one Jagdtiger by another, this would have almost certainly resulted in the breakdown of the one doing the towing, however in battle conditions there were many instances of this taking place! The recovery vehicles of the Support Company were to be used.

The towing trials conducted by Waffen Pruef 6 indicated the need for a central pull at the rear of the Tiger II/Jagdtiger, there were over a dozen devices proposed, one was fitted to Jagdtiger 305083, see Chapter 42.

24.4 Critical analysis of Jagdtiger on the battlefield

Advantages

The use of heavy armor with interlocking joints gave the vehicle great rigidity of construction which offered excellent protection in the event of a hit from an opposing anti - tank weapon at point blank range (50-100m).

The 12.8cm main armament was capable of destroying any opposing tank at a range of up to 3,000m. The size of the fighting compartment afforded relative comfort and room to maneuver allowing the crew to successfully carry out their functions with some speed. The accuracy of the gun, the clarity of the gun sight and precision movement of the aiming mechanism made targets at 2,000-3,000m range easy to destroy.

Because of the power of the main armament and the strength of the armor protection, this greatly boosted the morale of the crew, giving them the confidence they needed to take on the opposing armor.

In its intended/designated role, a well dug-in Jagdtiger was an extremely formidable weapon. Many opposing tanks were lost trying to assault such a vehicle from the front and flanks.

Disadvantages

The use of a separate projectile and cartridge case caused a slight delay in the loading the 12.8cm cannon, giving the Jagdtiger a slow rate of fire compared to vehicles firing single projectiles (a disadvantage in close combat conditions). In addition after firing a round it was necessary to bring the gun back to zero degrees elevation to allow the spent cartridge to eject properly and this incurred a further delay in the firing cycle. Due to the weight of Saukopf (pigs-head mantle), and to protect the delicate aiming mechanism an internal clamp was provided and an external frame travel-lock on the nose plate with which to clamp the gun during traveling. Both had to be released before combat.

As with all tank destroyers with fixed superstructures, only a limited traverse of the gun was available - 10 degrees left or 10 degrees right. If a target moved outside this range, the whole vehicle had to be slewed on its tracks. For this reason, any severe mechanical failure left the Jagdtiger wide open to flank attacks.

Several problems were highlighted because of the vehicle's excessive weight:

- Bridge crossing problems - routes had to be chosen carefully.
- A high rate of fuel consumption at a time of scarcity.
- Great stress on mechanical components.
- Low power to weight ratio (9.8 bhp/ton).

The early Jagdtigers suffered major transmission problems after 250-400 km driving.

[1]Between 25 December 1944 and 22 February 1945, one of the Porsche Jagdtigers was removed from the inventory of s.Pz.Jg.Abt.653. This is not recorded as destroyed in any of the official reports.
[2]36 Henschel Jagdtigers are recorded as issued to s.Pz.Jg.Abt.653 between 6 October 1944 and 13 January 1945.
[3]Dispatched to s.Pz.Jg.Abt 512 but did not get through they may have been redeployed to s.SS-Pz.Abt 501in May 1945! See Chapter 40/7
[4]Not substantiated by documents. See Chapter 40/8-9

Plate 196 - 197. Close up of the towing shackles, which could be fitted at each corner of the hull (Walter J. Spielberger).

25

Camouflage and Markings

25.1 Camouflage

This is a broad outline of the camouflage systems used by Jagdtiger units. Individual color schemes for each vehicle are indicated in the following photographic section.

Colors used:

Dunkel Gelb - a sand color that was the standard factory applied (RAL 7028) finish

Olive Grun - an olive green color formerly used by the Luftwaffe (RAL 6003)

Rot Braun - a red brown, chestnut color (RAL 8017)

Each vehicle released Prior to September 1944, from the factory with an overall Dunkel Gelb paint coating, they were issued with a 2kg tin of each of the three colors. Only the first four Jagdtigers released to combat units had this paint coating. The paints were very thick, similar to shoe polish, and thinning achieved using either gasoline or water; color strength was dependent on amount of thinning used. Olive Grun ranged from a light pea green color to almost black, while the Rot Braun from brick red to a deep maroon color. The tone of the colors was affected by the method of application. Occasionally, vehicles carried a spray gun, worked from a compressor run from the Jagdtiger's engine. Paints were also hand-applied using rags or brushes. This method gave a hard edge to the colors. Color schemes were usually chosen at platoon level.

With water used for thinning, the paint was unstable and even light rain would remove the colors or cause them to run. Gasoline was required for a durable finish, but unfortunately fuel was in very short supply. Mud, oil, dust and gasoline stains also affected the colors and they often proved as effective as the paints themselves.

After September 1944, the Jagdtigers were camouflage painted at the factory. No 305010 was the first to be released given a sprayed Three-Color scheme! The paint schemes used are illustrated in the color section, and notes on color schemes are contained in the photographic captions.

Because of the constant threat from Allied aircraft, good camouflage was absolutely vital. Local foliage was used as much as possible; this being held in place by wires rigged around prominent external features. Open country was to be avoided and vehicles parked in wooded areas for concealment during daylight hours.

White sheets were used during the winter snow months, to hide the tank destroyers in the heavy snowdrifts.

Both s.Pz.Jg.Abt 653 and s.Pz.Jg.Abt 512 used camouflage netting on a small number of vehicles.

25.2 Markings

a) National insignia: The Balkan Cross, the standard National Insignia on armored vehicles, was carried on the Jagdtiger these being factory applied. There were 4 such crosses sprayed on the fighting compartment, one on each side and two on the back plate either side of the double doors. Some vehicles only carried the right hand cross. These crosses were 0.15m x 0.15m, the central cross being black with line thickness 0.05m. White segments, the same width, completed the cross. With card masks used when spraying the crosses onto vehicles. Crosses were not always complete. Sometimes the black inner cross was missing. (See following photographs).

b) Unit insignia: By the time Jagdtigers were used in action, unit insignias were no longer carried on vehicles. Therefore, no unit insignia was used either for schwere Panzer Jäger Abteilung 653 or for schwere Panzer Jäger Abteilung 512.

Photographs of some of the last Jagdtigers to see action, in Austria (May 1945), show vehicles carrying individual motifs, such as a toy bear or the inscription 'Sunny Boy` on the left side of the front fighting compartment plate and gun-mantle.

c) Tactical numbering: Schwere Panzer Jäger Abteilung 653's Jagdtigers carried black and white tactical numbers on the side of the fighting compartment below the Balkan Cross. Each vehicle in a Company was assigned a three-digit number. The first digit indicated the Company, the second the Platoon and the third the position of the vehicle within that Platoon.

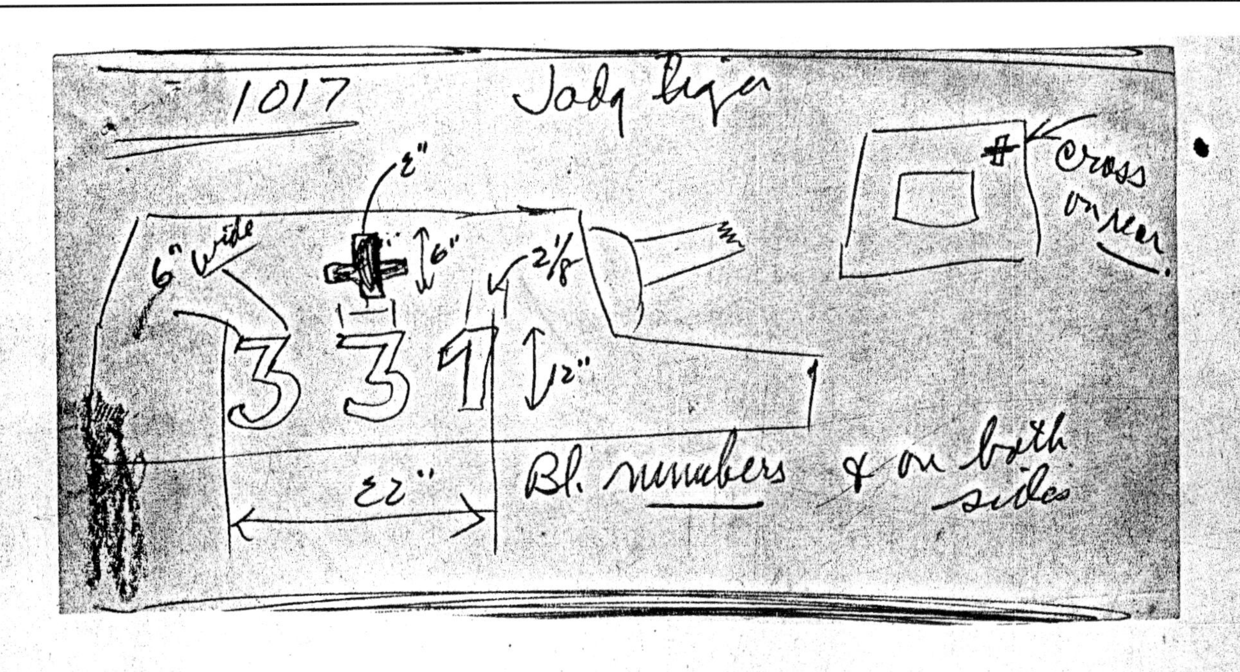

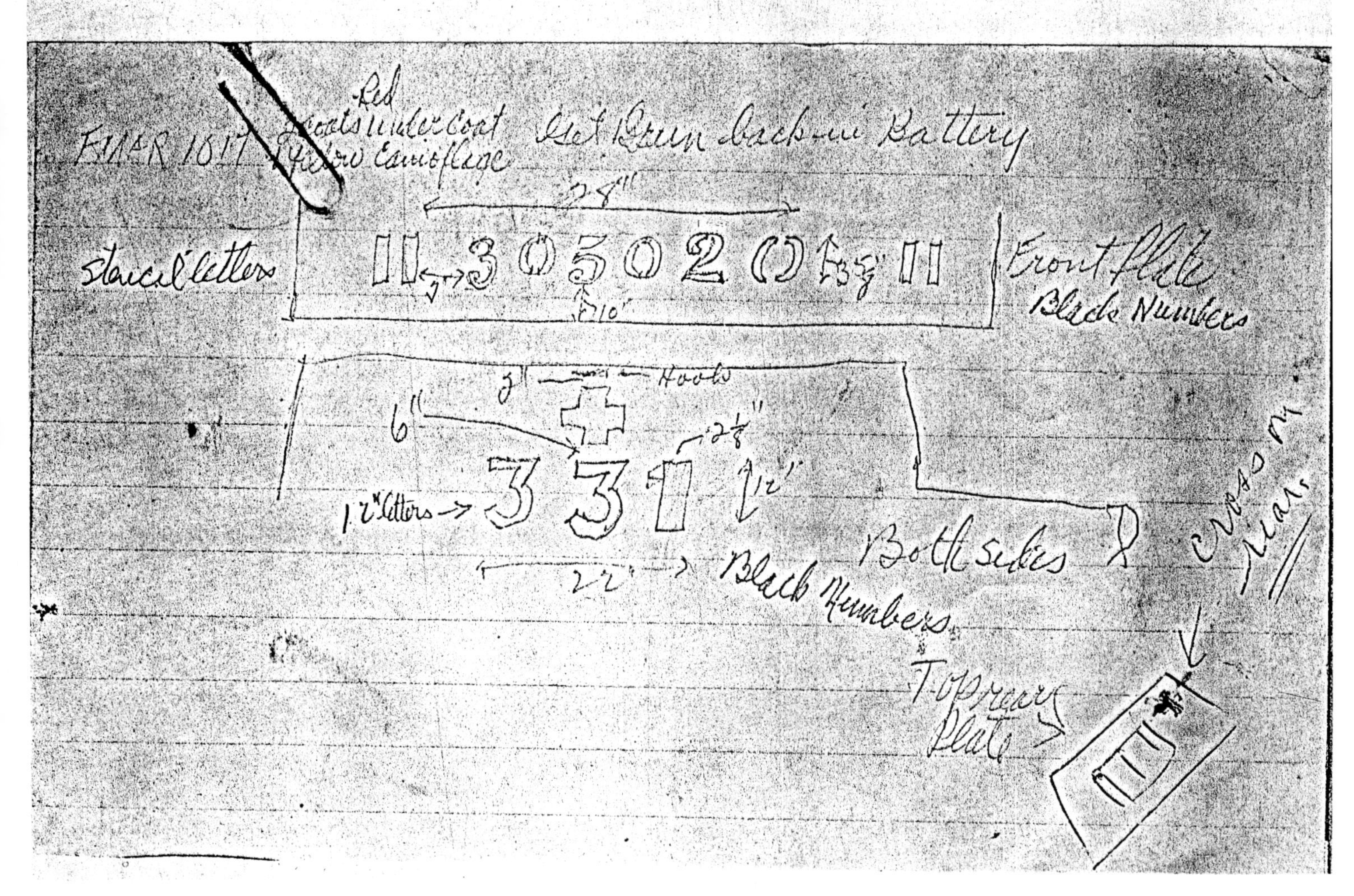

Notes on markings on Jagdtiger 331, from Intelligence Officer's notebook (APG).

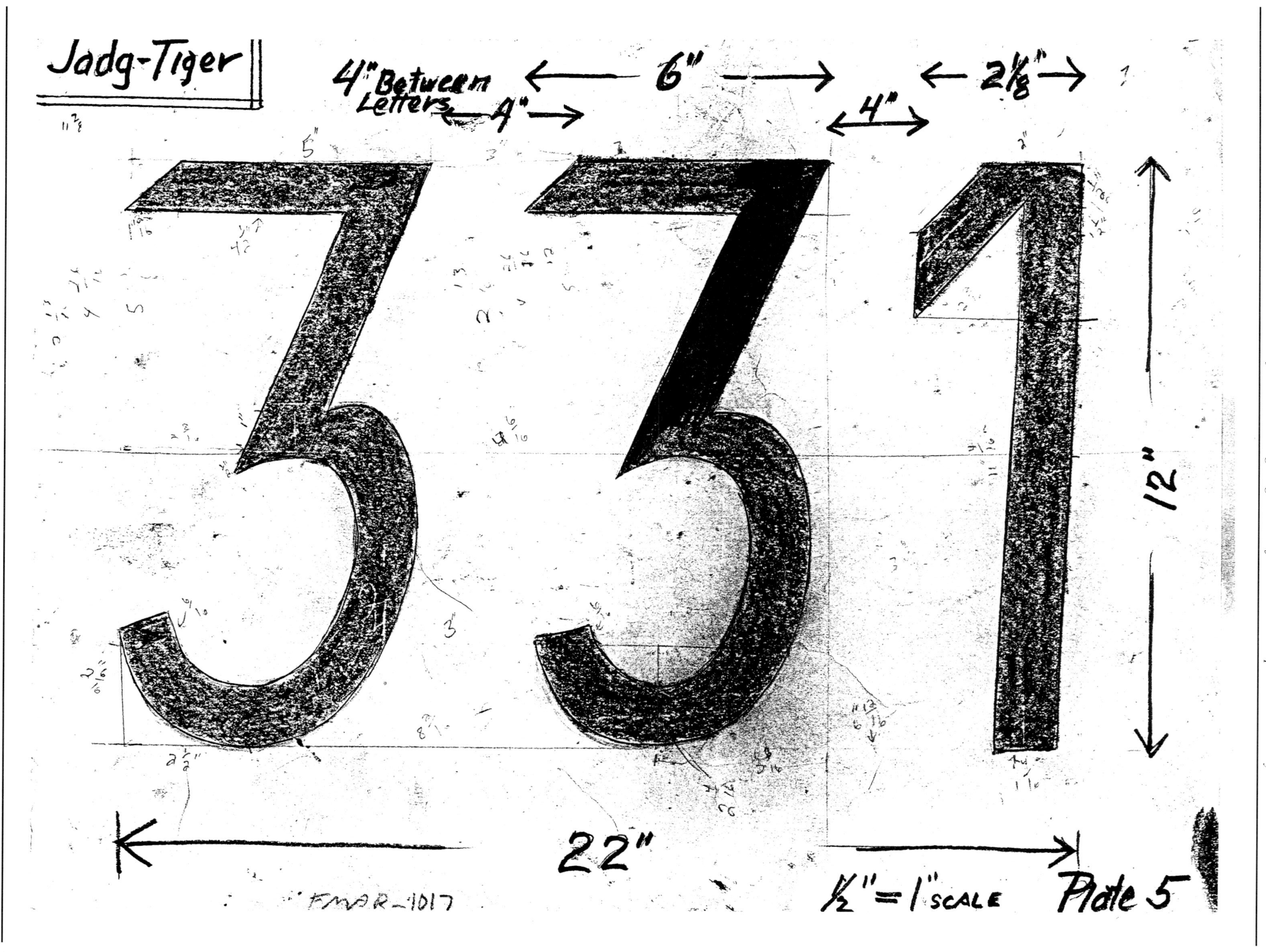
Jadg-Tiger
4" Between Letters
4"
6"
4"
2 1/8"
331
12"
22"
1/2" = 1" SCALE
Plate 5

Color Plate 1. Ferdinand recovery vehicle and Ferdinand, both are finished in Dunkel Gelb and Olive Grun Summer 1943. (Author)

Some examples:

332 = Third Company, Third Platoon, Second Vehicle in Platoon
314 = Third Company, First Platoon, Fourth Vehicle in Platoon
102 = First Company, O for HQ Vehicle, Second Vehicle

There was a maximum of 4 Jagdtigers in a Platoon. Numbers were sprayed on using card masks and were 0.3m high, with black figures 0.06 - 0.08m wide. Some had a white edge 0.01m wide.

Color Plate 2. Tiger II tactical number 233 with Zimmerit coating and sprayed three color scheme in Budapest. (Author)

Some of the 1/schwere Panzer Jäger Abteilung 512 vehicles are photographed carrying tactical numbers X1, X5 and X7. Others will have had similar numbers, but cannot be seen on photographs due to heavy foliage covering. One vehicle thought to be from 3/schwere Panzer Jäger Abteilung 512 is photographed carrying a "Y" as a tactical mark, see later.

d) Chassis numbers: These were stenciled in various places on the front glacis plate, black paint was used with the numbers being 85mm in height. They were often over painted with camouflage colors.

Color Plate 3. Fictional view of two Jagdtigers from the s.Pz.Jg.Abt.653, in the three colored camouflage scheme supporting coniferous foliage, Alsace area February 1945. (Author)

Color Plate 4. Jagdtiger 314 from the 3/s.Pz.Jg.Abt.653, painted in the "Ambush" color scheme, this was released from the factory on 8th November 1944. The Balkan cross has been factory applied in white outline only, Morsbronn March 1945. (Author)

Color Plate 5. The remains of Jagdtiger 332 in three color scheme. (Author)

Color Plate 6. Jagdtiger from the 2/s.Pz.Jg.Abt.653, the Ambush scheme can barely be seen underneath the camouflage netting, only the presence of the middle row of spare track links identifies this as a vehicle released in December 1944, Bruchsal area 27th March 1945. (Author)

Color Plate 7. The burned out remains of Jagdtiger 123 from the 1/s.Pz.Jg.Abt.653, it has a well recognizable Ambush camouflage scheme, Eppingen April 1945. (Author)

Color Plate 8. Front view of Jagdtiger tactical number X7 it is covered in dust which obscures its camouflage colors, it carries a small amount of coniferous camouflage. (Author)

Color Plate 9. Heavily covered in men and branches the color scheme is not obvious on this Jagtiger. (Author)

Color Plate 10. Again heavily camouflaged the Ambush paint scheme is just visible in a few areas. (Author)

Color Plate 11. The effect of men and branches certainly changes the appearance of these Jagdtigers. (Author)

Color Plate 12. Jagdtiger in heavy coating of Olive Grun with no attached foliage. (Author)

Color Plate 13. Jagdtiger "Sunny Boy," netted and under a tree is difficult to spot. (Author)

26

Schwere Panzerjäger Abteilung 653

26.1 History: Ferdinand phase

Schwere Panzer Jäger Abteilung 653 was formed in Brück Lethia in March 1943, out of the Sturmgeschutz Abteilung 197. It reached full strength with 45 Ferdinand Tank Destroyers in June 1943 - its commander was Major Steinwachs.

It was sent to the Eastern Front in June 1943 and assigned along with its sister battalion, schwere Panzer Jäger Abteilung 654, under Major Noak. The two, together with Sturmpanzer Abteilung 216 under Major Kahl, formed the Panzer Jäger Regiment 656, commanded by Oberstleutnant (Lt. Col.) Ernst Von Jungenfeldt, who had been appointed commander on 8 June 1943 in St Polten (Austria).

The unit designations at the time were:

I/656 (s.Pz.Jg.Abt.653) Major Steinwachs (45 Ferdinands)

II/656 (s.Pz.Jg.Abt.654) Major Noak (44 Ferdinands)

III/656 (Sturmpanzer Abt. 216) Major Kahl (42 Brumbars)

For the Kursk Offensive in July 1943, Panzer Jäger Regiment 656 was on the German northern flank, under XXXXI Corps in support of the 86th Infantry Division, reaching Alexandrovka in support of 292nd Infantry Regiment. A Wehrmacht report, dated 6 August 1943, stated that:

By 27 July 1943, 656 Regiment had accounted for 502 Soviet tanks, 100 artillery and 20 anti-tank guns.

After operation CITADEL was called off by Hitler on 13 July 1943, 653 alone had destroyed 320 Soviet tanks with the loss of 13 Ferdinands with 24 crew killed or missing.

By 1 August, 656 Regiment was down to 50 Ferdinands. They had been fighting against the huge Soviet offensive directed at the Orel salient. On 26 August the remaining Ferdinands and crews of 654 Abt were transferred to 653 Abt.

Regiment 656 continued fighting on the retreat from September to November. Generally, only about 20% of the Ferdinands were operational in the fighting at the Nikopol bridgehead.

In November, 656's total count went up to the destruction of 582 tanks, 344 anti-tank guns, 133 artillery and 103 tank destroyers. There were now only 42 remaining Ferdinands.

Plate 198. Ferdinand of the 1/s.Pz.Jg.Abt 653 in deployment for the start of "Operations Citadel" (Karlheinz Münch).

Plate 199. Ferdinand of s.Pz.Jg.Abt 653 ready to move into the attack on the northern flank of the Kursk offensive (Karlheinz Münch).

Plate 200. Ferdinand taking on ammunition, on 3 July 1943 (Karlheinz Münch).

Plate 201. Close up of the Ferdinand (Karlheinz Münch).

On 16 December 1943, an order 1/10027, stated:

The s.Pz.Jg.Abt.653 and Sturmpanzer Abteilung 216 were transferred to Army control, the Panzer Jäger Regiment 656 staff were canceled, only 4 of the 42 Ferdinands were operational. The Regiment staff of 656 became the staff of Panzer Brigade 101.

At this stage, the Ferdinands were returned to Nibelungen Werk for repair and modifications from January to March 1944. Major Steinwachs was recalled and handed over command to Major Baumungk.

In February 1944, 1/653 were sent to Italy with 11 out of a planned 14 Ferdinands. The other 3 remained in Nibelungen Werk. Their first action was against the Anzio beachhead along with s.Pz.Abt.508. Oblt. Ulbricht commanded the 1/653.

The first time the name "Elefant" was used was in a war diary ADK 14 (Italian), on 19 May 1944. At this stage there were nine remaining Elefants.

By 25 June 1944, 1/653 were down to the last two Elefants operational with one in repair, most had been lost to mechanical breakdowns in the unfavorable mountainous terrain.

The next day they were ordered to join the rest of the Battalion on the Eastern Front.

After the Nibelungen Werk refit, the bulk of the 653 Abt were returned to the Eastern Front. 2/653 and 3/653 were sent by rail

Plate 202. Ferdinand tactical number 134 with Ladungstrager B IV of the FKI. Kompanie 314 – which was also assigned to s.Pz.Jg. Regiment 656 (Karlheinz Münch).

Plate 203 - 207. Ferdinand No 714 of the s.Pz.Jg.Abt 654 in camouflage position before the Kursk offensive (ECPA).

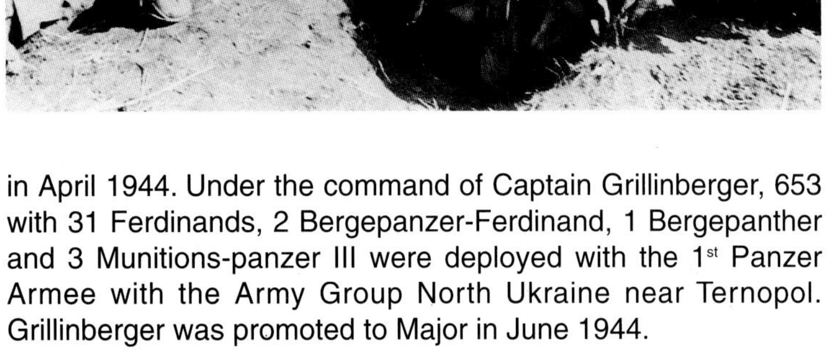

in April 1944. Under the command of Captain Grillinberger, 653 with 31 Ferdinands, 2 Bergepanzer-Ferdinand, 1 Bergepanther and 3 Munitions-panzer III were deployed with the 1st Panzer Armee with the Army Group North Ukraine near Ternopol. Grillinberger was promoted to Major in June 1944.

On 1 July, 653 had 34 Elefants (29 were operational), 1 Command Tiger (P), 1 Panther with Pz. IV turret, 2 Berge Panthers, and 2 Berge Tigers (P).

On 18 July 1944, the two companies were in action against the southern flank of the large Soviet offensive in the Ukraine,

Plate 208. Ferdinand moving into action (ECPA).

Plate 209. Brumbar in a battle exercise, on 3 July 1943. Major Kahl Commander, and Lt. Stemann Adjutant of Sturm Pz.Abt 216 (Karlheinz Münch).

which quickly broke through the German lines. After two weeks of vicious fighting 653 were down to less than company strength with Elefant, See Chapter 27.

26.2 s.Pz.Jg.Abt.653 Jagdtiger phase

On the 20 June 1944, a Panzer Jäger replacement and training Battalion were formed in Munchen to train part of s.Pz.Jg.Abt. 653 with the latest and heaviest Panzer Jäger vehicle developed for the tank destroyers, this vehicle being the Jagdtiger. One Jagdtiger, 305003, was made available, in July, for the course.

The s.Pz.Jg.Abt. 653 were chosen for Jagdtiger operations because of their combat experience with Ferdinand/Elefant operations in Russia and Italy. Great value was placed on the Jagdtigers and much was expected of them, a secret weapon to help win the war! At almost weekly intervals "Major General Thomale", the Inspector General of Tank Troops, and OKH would review the s.Pz.Jg.Abt. 653 unit.

The official War Diary of s.Pz.Jg.Abt.653 did not survive the War, only the reports sent to OKH/Inspector General of Tank Troops. The following section has been pieced together with help from Tom Jentz and enormous help from Karlheinz Münch, who has been researching the Unit since 1986 - he has contacted

Plate 210. Brumbar moves to the front (Karlheinz Münch).

over 300 former personnel, his book on the history of the 653 battalion was published in 1998.

It is not practical to allocate crews to individual vehicles. The Jagdtiger was so unreliable that crews were constantly changing

Plate 211 - 212. Brumbar on the front line (Karlheinz Münch).

65

Kriegsgliederung

der schw. Pz. Jäg. Abt. 653 (Elefant)

Stand: 1. Juli 1944

Gefechtsstab:

Wirtschaftlich auf Stabskompanie angewiesen:

Kdr. Major Grillenberger — schw. 653 — Adj. Oblt. Scherer

Gepäck-Troß | Gruppe Führer: gep. Sankra; Fla-Vierling auf Panther; Fu-Stelle im Spw.; Bef.-Panther mit Pz. IV-Turm; Bef.-Tiger

Führer: Lt. Klos — Stabskompanie

Nachrichten-Zug | Fla-Zug: 2cm Fla-Vierling; 2cm Fla-Vierling auf T34; 2cm Fla-Vierling auf Sfl. | Pi-Zug | Erkundungszug: Spw. | Pz.-Jäger Zug: Muni-Pz. T34; Elefant; Elefant; Elefant; Elefant; Elefant; Elefant

Gepäck-Troß | Verpflegungs-Troß | Staffel für Verwaltung u. Nachschub | Sani-Trupp | Gefechts-Troß | Kfz-J-Staffel

Nachschubtransportraum	
Soll	38 to
Jst	20 to
Einsatzber.	20 to

Führer: Oblt. Kretschmer — 3. Kompanie

Gruppe Führer: Elefant; Elefant

3. Zug: Elefant; Elefant; Elefant; Elefant | 2. Zug: Elefant; Elefant; Elefant; Elefant | 1. Zug: Elefant; Elefant; Elefant; Elefant

Gepäcktroß | Gefechtstroß II | Gefechtstroß I: Muni-Pz III; Muni-Pz. 34; Berge-Elefant | Kfz.-J-Staffel

Führer: Oblt. Salamon — 2. Kompanie

Gruppe Führer: Elefant; Elefant

3. Zug: Elefant; Elefant; Elefant; Elefant | 2. Zug: Elefant; Elefant; Elefant; Elefant | 1. Zug: Elefant; Elefant; Elefant; Elefant

Gepäcktroß | Gefechtstroß II | Gefechtstroß I: Muni-Pz-III; Muni-Pz.-III; Berge-Elefant | Kfz.-J-Staffel

Führer: Lt. Demleitner — Werkstattkompanie

Troß | Nachrichtengerätwerkst. | Waffenmeisterei | 3. Zug (Bergezug) | 2. Werkstattzug | Gruppe Führer

Benennung:	Offz.	Beamte	Uffz.	Mann	Hiwis	Gesamt	Karab.	Pist.	M.Pi	Armee-Pi(b) als M.Pi. Ers.	Leucht.Pi	M.G. 34 im Pzw.	2cm Fla-vierling 38	8,8cm Pak 43/2 L71 f. Elefant	8,8cm Kwk 36 L/56 f. Tiger	7,5cm Kwk 40 L/48 f. Panther
Soll	22	8	235	731	–	996	639	399	101	–	58	64	4	34	1	1
Jst	19	8	196	756	47	1026	633	362	82	35	36	54	4	34	1	1
Fehl	3	–	39	–	–	–	6	37	19	–	22	10	–	–	–	–

Benennung:	m. Krad	s. Krad m. Bwg	Pkw.	Lkw.	Maultiere m.	Maultiere s.	Zgkw.	Spezialgerät als Sd. Anh.	Panzer-Jäg „Elefant"	Berge-Pz. „Elefant"	Panzer VI „Tiger"	Panzer V „Panther"	Berge-Pz. „Panther"	Muni-Pz. III	Muni-Pz. T34	Sfl. T34 f. Fl.-vierl.	gep. Sankra
Soll	26	9	45	90	11	–	29	11	34	2	1	–	2	4	–	–	1
Jst	23	6	39	56	25	5	23	13	34	2	1	2	–	3	2	1	1
Fehl	3	3	6	34	siehe Bemerk.		6	–	–	–	–	1 Berge Pz zum Fla.-Pz. 1 B.Pz. zum Bef.Pz. umgebaut		2 Beulepanzer T34 als Muni-Pz. umgebaut		–	–

Bemerkung:

A) 1. Kompanie u. 1. Werkstattzug befinden sich im Einsatz in Jtalien

B) Als Ausgleich für die fehlende 1. Kompanie ist von Stabskp. der Pz.Jäg.-Zug u. Erkundungs-Zug zur Kampfgruppe „Wiesenfarth" zusammengefaßt.

C) Es fehlen 29 Lkw wofür 19 Maultiere vorhanden sind. Das sich ergebende Fehl an Tonnageraum beträgt 76,5 to.

Plate 213. Jagdtiger chassis numbers 305006, 305007 and 305008 in Fallingbostel goods yard, in September 1944 (Wolfgang Schneider).

vehicles. Where crews were known to be with a particular Jagdtiger in a time period, these are indicated.

The unit received its first Jagdtigers in September/October 1944. By March 1945, they had taken delivery of 42 Jagdtigers, which formed the 3 fighting companies (each equipped with 14 vehicles) of the unit.

See the documents included for the task organization of s.Pz.Jg.Abt. 653.

For simplicity, I will indicate the strength build-up, training, deployment and combat of this unit month by month, until its surrender in Austria on the 7 May 1945. Production and delivery figures are also given, month by month, as these obviously had a direct effect on the battalion's combat readiness and effectiveness. At the same time, I shall dispel some of the rumors and myths associated with the Jagdtiger.

Where tactical numbers can be linked to chassis numbers, these are shown, e.g., 305023 (113). There was no logical number-sequence linking the two!

Note: More historical accuracy should be attributed to the official records compiled from the dates which are printed in italics, rather than the veterans' accounts presented, even where personal diaries have been kept!

26.3 Jagdtiger deliveries before September 1944

Deliveries to	Transported	No	Type	Chassis No
Kummersdorf	*April*	*2*	*(P)*	*305001*
			(H)	*305002*
Mielau	*30 June 1944*	*1*	*(P)*	*305003*
Wa Pruef 6	*August*	*1*	*(P)*	*305004* [1]
Mielau	*28 August*	*3*	*(P)*	*305006*
				305007
				305008

26.4 Jagdtigers built by Nibelungen Werk before September 1944

				Notes
February 1944	*(2)*	*305001*	*(P)*	No stowage hooks or
		305002	*(H)*	Gun supports - No Zimmerit.
July 1944	*(3)*	*305003*	*(P)*	
		305004	*(P)*	All with Zimmerit.
		305005	*(P)*	
August 1944	*(3)*	*305006*	*(P)*	First with gunlock.
		305007	*(P)*	
		305008	*(P)*	

26.5 Hulls built by Eisen Werk before September 1944.

November	*1943*	*1*	
December	*1943*	*3*	
January	*1944*	*5*	
February	*1944*	*8*	
March	*1944*	*10*	
April	*1944*	*15*	
May	*1944*	*12*	
June	*1944*	*10*	
July	*1944*	*7*	*Air raid 25 July 1944*
August	*1944*	*zero*	Eisen Werk heavily hit; 2000 killed!

[1]This Jagdtiger was issued to Wa Pruef 6 before 1 September 1944.

27

Schwere Panzerjäger Abteilung 653 September 1944

27.1 September 1944

At the beginning of September 1944, what remained of the Elefant Abteilung were undergoing repairs in the Krakau area of Poland.

Previously, they had been involved in bitter fighting near Ternopol against the great Soviet offensive, which started on, 18 July 1944. On the first day of the attack, their role was to provide rearguard cover for the German withdrawal from the area. They were able to inflict heavy casualties on the Soviets but had little chance of retrieving the bulk of their heavy equipment from their precarious positions, where when eventually over-run, 19 of the 33 Elefants were lost, at great cost to the Soviet tank forces. Further fighting took place until, on 3 August 1944, they were pulled out of the front-line with only 12 Elefants remaining, all needing repair, and the 2 Bergepanzer Elefants were also lost. The 653 had suffered surprisingly low casualties with only one Elefant crew being killed by a direct hit, 20 others had received various bullet and shrapnel wounds. Their refit area was near Krakau. At Krakau, they were joined by the remnants of 1/653 that had been ordered back from Italy with their last two Elefants.

The following assessment of the July operations was sent to Heeres Gruppe Nordukraine:

27.2 Report dated 1/8/1944

1. /Panzerjag - Regt. 656 (schw.Panzerjag.Abt.653)
Panz. Lehrkommando b/Heeresgr. Nordukraine

1. Personnel log, 1/7/44 to 31/7/44	*a. Total 996 personnel, 22 officers, 235 sub officers, and 731 men.* *b. Casualties and departure, 5 dead, 20 wounded, 3 ill.*
Material log,	*Elefant (quota 31, operational 0, short-term repair 2, long repair 10).* *Berge panzer Elefant (quota 2, on strength 0)* *Berge Panther (quota 2, require short term repair 2)* *Flak Pz IV (quota 4, operational 4).*

Plate 214. Jagdtiger chassis number 305007 is prepared for rail transport (Wolfgang Schneider).

Nordukraine   Nikolaj 34   07

I./Panzerjäger-Regt. 655

Meldung vom 1.8.44.....   Verband: (schw. Panzerjäg. Abt. 653)

Unterstellungsverhältnis: **Panz. Lehrkommando b/Heeresgr. Nordukraine**

1. Personelle Lage am Stichtag der Meldung:

a) Personal:

	Soll	Fehl
Offz.	22 + 8 Bea	2
Uffz.	235	41
Mannsch.	731	-
Hiwi	-	-
Insgesamt	996	-

c) In der Berichtszeit eingetroffener Ersatz:

	Ersatz	Genesene
Offiziere	-	-
Uffz. u. Mannsch.	24	10

b) Verluste u. sonstige Abgänge in der Berichtszeit vom 1.7...bis 31.7.44......

	tot	verw	verm	krank	**vers.**
Offz.	-	1	-	1	1
Uffz. u. Mannsch.	5	19	-	2	-
Insges.	5	20	-	3	1

3 Uffz., 8 Mannsch. u. 4 Freiw. sind <u>bisher</u> nicht zur Einheit zurückgekehrt

d) Über 1 Jahr nicht beurlaubt:

insgesamt Köpfe / % d. Iststärke

davon:	12-18 Monate	19-24 Monate	über 24 Monate
	8	-	-

Platzkarten im Berichtsmonat zugew.: -

2.) Materielle Lage:

		Gepanzerte Fahrzeuge						Kraftfahrzeuge			
		Berge-Pz. Wg. Panther	~~Ferdinand~~ **Elefant**	Pz. III	Mun. Pz. III	Kr Pz Wg	Pak Sfl ~~Ferdinand~~ **Elefant**	Kräder m. ang. Bwg.	Kräder sonst	Pkw gl.	Pkw (o)
Soll	zahlenm.	2	2	-	4	1	31	9	26	43	2
einsatzber.	zahlenm.	-	-	-	-	1	-	1	13	12	2
einsatzber.	% d. Solls	-	-	-	-	100	-	11	50	28	100
In kurzfr. Instands. b. 3 Wochen	zahlenm.	2	-	-	-	-	2	5	10	22	-
In kurzfr. Instands. b. 3 Wochen	% d. Solls	100	-	-	-	-	7	55	38	52	-

		Noch Kraftfahrzeuge						Waffen			
		Maultier	Lkw. gel	Lkw. (o)	Lkw. to	Kett. Kfz. Zgkw.	Kett. Kfz. RSO	Vierl. Flak	s. Pak 8,8cm	M.G.	sonst. Waffen
Soll	zahlenm.	11	90		291	29	-	4	31	52	-
einsatzber.	zahlenm.	-	10	12	585	1	-	4	12	37	1 KwK 7,5
einsatzber.	% d. Solls	-	24		21	35	-	100	39	71	-
in kurzfr. Instands. b. 3 Wochen	zahlenm.	26	9	24	935	22	-	-	1	-	-
in kurzfr. Instands. b. 3 Wochen	% d. Solls	236	36		32	77	-	-	3	-	-

Bei sämtlichen Zahlenangaben ist die in Italien eingesetzte 1. Komp. außer Betracht geblieben.

Kriegsgliederung

Stand 5. August 1944 **68**

der schw. Pz. Jag. Abt. 653 (Elefant)

Gefechtsstab:
Wirtschaftlich auf Stabskompanie angewiesen!

Kdr. Major Grillenberger — Adj. Oblt. Scherer

Gepäck-Troß	Gruppe Führer.
	gep. Sankra · Fla Vierling auf Panther · Fu-Stelle im Spw · Bef.-Panther mit Pz IV-Turm

Führer: Lt. Wiesenfarth — **Stabskompanie**

Nachrichten-Zug	Fla-Zug (2cm Fla-Vierling · 2cm Fla-Vierling auf T34 · 2cm Fla-Vierling auf Sfl.)	Pi-Zug	Erkundungszug	Panzer-Jäger-Zug

Gepäck-Troß	Verpflegungs-Troß	Staffel für Verwaltung u. Nachschub	Sani-Trupp	Gefechts-Troß	Kfz.-J.-Staffel

Nachschubtransportraum	
Soll	38 to
Ist	20 to
Fehl	18 to

Führer: Oblt. Kretschmer — **3. Kompanie**

Gruppe Führer (Elefant, Elefant)

3. Zug	2. Zug	1. Zug
Elefant, Elefant, Elefant, Elefant	Elefant, Elefant, Elefant, Elefant	Elefant, Elefant

Gepäcktroß	Gefechtstroß II	Gefechtstroß I	Kfz.-J.-Staffel

Führer: Oblt. Salamon — **2. Kompanie**

Gruppe Führer

3. Zug	2. Zug	1. Zug

Gepäcktroß	Gefechtstroß II	Gefechtstroß I	Kfz.-J.-Staffel

Führer: Lt. Demleitner — **Werkstattkompanie**

Troß	Nachrichtengerätwerkst.	Waffenmeisterei	3. Zug (Bergezug)	2. Werkstattzug	Gruppe Führer

Benennung:	Offz.	Beamte	Uffz.	Mann	Hiwis	Gesamt	Karab.	Pist.	M.Pi.	Armee Pistol als M.P. Ers.	Leucht. Pi.	M.G. 34 im Pz.W.	2cm Fla Vierling 38	8,8cm Pak 43/2 L/71 f. Elefant	7,5cm Kwk 40 L/48 f. Panther
Soll	22	8	235	731	–	996	647	359	127	–	41	52	4	31	1
Ist	20	7	194	768	19	1008	613	297	48	35	7	37	4	12	1
Fehl	2	–	41	–	–	–	34	62	44	–	34	15	–	19	–

Benennung:	m. Krad	s. Krad m. Bwg.	Pkw.	Lkw.	Maultiere m	Maultiere s	Zgkw.	Spezialger. als Sd. Anh.	Panzer Jäg. „Elefant"	Berge-Pz „Elefant"	Panzer VI „Panther"	Berge-Pz „Panther"	Muni-Pz. III	Sfl T-34 f. Fla Vierling	gep. Sankra
Soll	26	9	45	91	11		29	11	31	2	–	2	4	–	1
Ist	23	6	36	55	22	4	23	13	12	–	2	–	–	1	1
Fehl	3	3	9	36	siehe Bemerk.		6	–	19	2	1 Pz zum Fla Pz. 1 zum Bef Pz. umgebaut		4	–	–

Bemerkung:

A) 1 Kompanie und 1 Beamter, 9 Uffz., 75 Mann, 5 Hiwis von Stabskp. u. Werkstattkp. befinden sich im Italieneinsatz. Die Männer von St.Kp. u. Werkst.-Kp. sind im angegebenen Soll u. Ist enthalten. 1 Kp. nicht.

B) Es fehlen 29 Lkw. wofür 15 Maultiere vorhanden sind. Der Laderaum ist dadurch nicht ausreichend ausgeglichen. Die restlichen 7 fehlenden Lkw. sind Spezialfahrzeuge

Short assessment by the Battalion Commander

Training: *Before the start of operations, there had been a continuous three month training period, training in each aspect is good. The previous experience against foreign troops was fully applied. The last major battle day fully confirmed all points. Difficulties encountered were because of the operation command issued, no one was familiar with the conventional system for tanks and assault guns. During the present repair period, the requirement is for three weeks renewed training activity for all non-tasked sections, separate requirements for officers, NCO's (commanders). Tank service without radio, special training (Radio, tank driving, repair service, engineers, antiaircraft troops, and recovery platoon). The majority of the unit must be reorganized with the supply of new and more reliable armor.*

Mood of the troops: *The troops fully understood the situation, the demand of the last operation, compared with the mechanical aspect of the vehicles, this made it impossible to stop the attack. They proved themselves by holding firm, but the outcome was obvious. The troops feel superfluous without armor, to take on pure infantry role. Here in this deployment area the technical working of the weapon was low with an unrecognizable consequential effect for the infantry. The troop has the endeavor, because of all the breakdowns, for getting ready with the new weapons training; we require a quick decision to be able to take on new role.*

Special difficulties: *The mistake of having one workshop platoon, the repair crane (special Kfz. 9/1) and 17 s Lkw, all supply vehicles were working by themselves carrying high levels of supplies and ammunition. This took enormous strength from the unit. Our mission was all rearguard resistance and delaying actions, which was impractical coupled with only one loading point. It was not possible for full recovery from our stretched out positions by one train, the panzer recovery was impracticable, especially when the plan was moderately changed because of enemy pressure which disrupted the main roads when the bridge was opened up. The repairs have been delayed through lack of success in supplying the replacement parts, due to train delays. As for re-armament, some have minimal experience with "Tiger"; as for as heavy Jagdpanzer (Jagdtiger) order to be received, but with availability, we must also have training time for drivers. Repair and service work, 3 - 4 weeks becomes justified. Other as-*

Plate 215. All these Jagdtigers had the early type of gun travel locks (Wolfgang Schneider).

Plate 216. Close up of Jagdtiger chassis number 305008 complete with fighting compartment cover and muzzle cover (Karlheinz Münch).

pects involving the re-arming, will require considerably more time (8 - 10 weeks) and the re-organization from base level, even while the repairs continue. There is no time to lose in getting 3 driving instructors and 3 training experts for repair and service work to stop any major delays from developing.

Mobility: *At present 80% of the Panzers require repair.*

Battle value and possible use: *After new supply of Panzers, repair of broken down Panzers and after supply of repair materials, for each operational opening we are suitable.*

Major and Battalion Commander.

27.3 The first training with the new Jagdtigers

With only 14 Elefants remaining, two of which were still in transit from Italy, s.Pz.Jg.Abt 653 were now down to company strength in heavy equipment. Most of the Elefants had been lost to mechanical failure and the inability to evacuate them to the single railhead.

In the last week in August, Elefant crews were sent by rail from Krakau to Fallingbostel in Northern Germany to join Training and Reserve Battalion 500. The journey took six days, with their arrival on 26 August. They were to attend a driver-training course, which took place between 9 and 20 September 1944. For this, three Porsche Jagdtigers were made available and which had been recorded as dispatched from Nibelungen Werk on 28 August 1944 for Ersatzheer (Mielau). They were chassis numbers 305006, 305007, 305008. The three Jagdtigers remained in Fallingbostel, until October.

Plate 217. Close up of chassis number 305006 with the missing spare track. The Zimmerit was applied up to the top row of spare track hooks (Karlheinz Münch).

After the driver training course the crews immediately transferred to the Linz area.

Under the supervision of Lt Demleitner, the Elefants were being prepared for operational readiness when, on 19 September, Major Grillinberger was ordered to deploy 653 back to the front line with 17 Army (Heeres Gruppe A). The Elefants now belonged to 2/653; command was taken over by Lt. Heinrich Teriete. Fortunately for the Elefant Company, the front-line in this area would remain stable for nearly 4 months, as all the Soviet offensive actions were at this time being directed against Army Group South, in Hungary.

27.4 Jagdtiger deliveries in September 1944.

Delivered to	Transported on	No	Type	Chassis No	Tactical No
Mielau	*28.8*	*3*	*(P)*	*305006*	?
			(P)	*305007*	?
			(P)	*305008*	?

Plate 218 - 220. (*above and p. 179*) Close up of early Jagdtigers (Karlheinz Münch).

27.5 Jagdtigers built by Nibelungen Werk in September 1944

The factory completed 8 Jagdtigers against 12 programmed in September.

Chassis No:		
305009	*(P)*	
305010	*(P)*	Last with Zimmerit, the "A" type gunlocks start.
305011	*(P)*	
305012	*(P)*	Last of the Porsche suspension Jagdtigers
305013	*(H)*	
305014	*(H)*	
305015	*(H)*	
305016	*(H)*	

On 9 September 1944 the order to Nibelungen Werk, "discontinue use of Zimmerit".

27.6 Hulls built in Eisen Werk in September 1944

There were 4 Hulls completed in September giving a running total of 75.

There was no bombing of the factories in September.

Note: Jagdtigers 305006, 305007, 305008 later taken on by s.Pz.Jg.Abt.653.

28

Schwere Panzerjäger Abteilung 653 October 1944

28.1 s.Pz.Jg.Abt 653 October 1944

In October 1944, s.Pz.Jg.Abt. 653, which had been chosen by Hitler himself because of their combat experience and success with the Ferdinand/Elefant vehicles, were the first unit to be equipped with the new Jagdtigers. The troops from s.Pz.Jg.Abt. 653 were ordered to transfer to Döllersheim near Zwettl, for reformation back to full Battalion strength.

In early October, they were deployed with 17 Army (Heers Gruppe A) in Poland east of Krakau. Their last 14 Elefants were ordered to be transferred, in situ, 2/s.Pz.Jg.Abt. 653 became s.Pz.Jg.Kp. 614 along with some of the former 653 personnel including some Ferdinand crew, the Works Company Commander Lt. Demleitner and his engineers. Major Grillenberger was to retain command of the 653 unit, Oblt Scherer remained as his Adjutant. They both traveled to Döllersheim, in early October, to set up a new Battalion HQ in a large wooden hut in the military camp.

From Döllersheim, Grillenberger would oversee the re-equipping of 653 and the Jagdtigers, in production, in Nibelungen Werk. He made several visits to the factory and met with Dr. Judtmann, the Managing Director of Nibelungen Werk.

On 16 October, a heavy Allied-bombing raid, dropping 143 tons of high explosive, was made on the Nibelungen Werk. This as usual had not been accurate and did not seriously damage the production line, and, consecutive chassis sequences continued. However, a delay was incurred clearing-up the superficial damage. Consequently, the month's production was only nine vehicles, just 45% of planned output. This would obviously have a direct effect on the equipping of 653 and the time required reaching combat readiness. Grillenberger inspected the damage for himself and which he reported directly to OKH.

During October, the Battalion's first Jagdtigers started to arrive at Battalion HQ. By the end of the month, twelve Jagdtigers

Plate 221. Side view of Jagdtiger chassis number 305009 at Döllersheim railhead, in October 1944. This was the first Jagdtiger delivered to the reformed s.Pz.Jg.Abt 653 (Karlheinz Münch).

Plate 222. Jagdtiger chassis number 305009 was the last to have the early type of gun travel lock (Karlheinz Münch).

Plate 223. Close up of the suspension of chassis number 305009 (Karlheinz Münch).

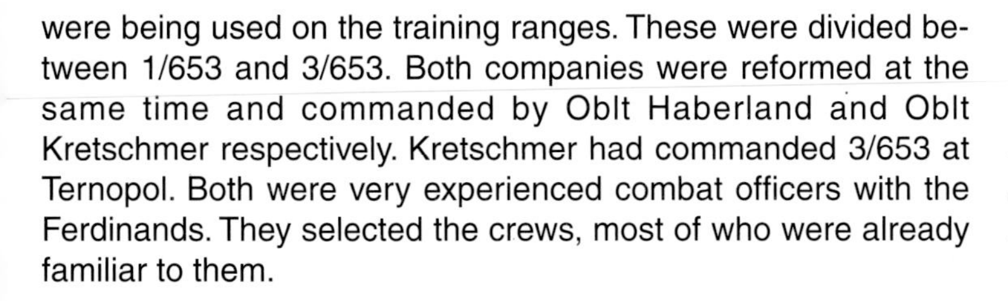

were being used on the training ranges. These were divided between 1/653 and 3/653. Both companies were reformed at the same time and commanded by Oblt Haberland and Oblt Kretschmer respectively. Kretschmer had commanded 3/653 at Ternopol. Both were very experienced combat officers with the Ferdinands. They selected the crews, most of who were already familiar to them.

Of the twelve vehicles delivered in October:

a) The first three (all Porsche type) 305009, 305010 and 305011 were actually assigned to the Ersatzheer (Army Reserve) and they were de-trained on 7 October 1944. These were later reassigned to s.Pz.Jg.Abt.653.

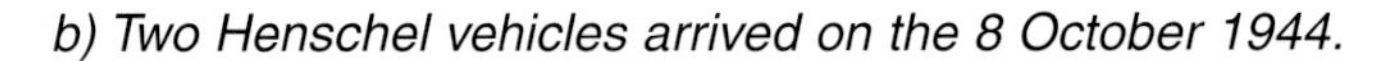

b) Two Henschel vehicles arrived on the 8 October 1944.

Plate 224 - 226. More views of chassis number 305009 (Karlheinz Münch).

Plate 227. A good view of Jagdtiger chassis number 305009 in October/November 1944 (Karlheinz Münch).

c) Two trains delivered four Henschel vehicles on the 25 October 1944.

d) The three Porsche vehicles were also sent back to Döllersheim from Fallingbostel in October.

The Jagdtigers were split between 1/653 and 3/653, with crews still awaiting vehicles being deployed to the factory to assist with the production operations.

The main purpose of the factory training was to gain experience with the Jagdtiger, which had a similar role to the Ferdinand, but was vastly different in design to their previous vehicle.

The Jagdtigers guns were sighted-in to the optics on the firing ranges at Döllersheim.(1) The ammunition was supplied from the factory in Magdeburg. The Panzer Jägers were highly impressed with both the accuracy and potency of their new Panzers, and they were very keen to take them into action.

28.2 Jagdtiger deliveries in October 1944

Delivered to	Transported on	No.	Type	Chassis No.	Tactical No.
Ersatzheer	*5-7/10/44*	*3*	*(P)*	*305009*	?
			(P)	*305010*	301
			(P)	*305011*	?
s.Pz.Jg.Abt.653	*6-8/10/44*	*2*	*(H)*	*305013*	?
			(H)	*305014*	?
Putlos	*14/10/44*	*1*	*(P)*	*305005* (2)	-
s.Pz.Jg.Abt.653	*23/10/44*	*1*	*(H)*	*305015*	?
s.Pz.Jg.Abt.653	*23/10/44*	*3*	*(H)*	*305016*	?
			(H)	*305017*	?
			(H)	*305018*	?

Plate 228. Jagdtiger chassis number 305010, it has the new type of gunlock (Karlheinz Münch).

28.3 Jagdtigers built in Nibelungen Werk in October 1944

The factory completed 9 Jagdtigers against 20 programmed in October.

Chassis No:305017
305018
305019
305020
305021
305022
305023
305024
305025

Note: bombing raid on 16 October 1944; 143 tons of high explosive, from photographs, 1 hull was damaged before it had been put on assembly line. It was, therefore, not allocated a chassis number and would have been scrapped!

28.4 Hulls built in Eisen Werk in October 1944

There were 13 Hulls completed in October giving a running total of 88.

(1)Firing trials did not start until 1/11/44, see Map 12, Chapter 40.7
(2)Jagdtiger 305005 was not to be used in combat because of its faulty armor.

29

Schwere Panzerjäger Abteilung 653 November 1944

29.1 November reformation

Throughout November, almost a full compliment of men had been assigned to the Battalion. It was, however, a different story as far as their heavy equipment was concerned.

With the availability of 12 Jagdtigers, driver training and gunnery practice started at the beginning of the month and lasted for just over two weeks. Jagdtiger 305011 (P) is recorded as one of the main driver training vehicles. Certificates of driver training were issued to the drivers on completion of the training course. All test firing was done on the Döllersheim ranges starting with training ammunition.

On 5 November, Jodl signed the order for the deception and secrecy plan for the forthcoming Ardennes offensive, and 653 were to be used as part of the northern flank defenses with their code name being HUBERTUS 1.

The original intention had been for both 1/653 and 3/653 to be combat ready by the end of November. In the event, only a further eight Jagdtigers had been delivered by then:

a) A Porsche vehicle, 305012, delivered on 10 November, is assigned to the Ersatzheer.

Plate 232. Battalion HQ, a wooden hut in Döllersheim in November 1944 (Karlheinz Münch).

Plate 229 - 231. Useful close ups of chassis number 305010 in its three-color camouflage scheme. This was the first Jagdtiger assigned to 2/s.Pz.Jg.Abt 653 for training, and was the last Jagdtiger with a coating of Zimmerit (Karlheinz Münch).

b) A Henschel vehicle received on 21 November.

c) Three Henschel vehicles received on 24 November.

d) Four Henschel vehicles received on 28 November. This last delivery had not been through test firing by the end of November.

To further compound the delay, the factory had had a very poor production period with only 30% of planned production achieved. Just six new Jagdtigers were built in the month. By then, it looked very unlikely that 1/653 and 3/653 would be combat ready by the end of December.

Organizational changes were effected in November. Oblt Schulte had replaced Lt. Demleitner as the works company commander. He and his men were trained on Jagdtiger maintenance and repair by working with the Nibelungen Werk construction engineers. New recovery vehicles, Berge Panthers, were to be used by the Support Company under the command of Hptm Ulbricht, two of these had been in service with the older organization. Hptm Konnack was appointed Stabskompanie commander; he replaced Lt. Klos who was wounded in July (Lt. Wiesenfarth had taken over for the intermediate period) until Konnack was appointed.

Grillinberger informed OKH, at the end of November, that two fighting companies would not be fully ready for the Ardennes offensive. Twenty-one Jagdtigers were in Döllersheim by the end of November. Planning for 2/653 was also taking place, in November. It was to be commanded by Oblt Wiesenfarth who after his good performance in Italy was keen to take on the new fighting company.

29.2 Organization staff positions s.Pz.Jg.Abt 653 30/11/44

Commander	Major Rudolf Grillinberger	
Adjutant	Oblt Kurt Scherer	
Ordonnanzoffizier	Lt. Herman Knack	
HQ Company	Hptm Bernard Konnak	
1 Company	Oblt. Werner Haberland	(101)
1 Platoon Commander Lt. Knippenberg		(111)
2 Platoon Commander Ofw Koss		(121)
3 Platoon Commander Ofw Kinnberger		(131)
2 Company	Oblt. Robert Wiesenfarth	(201)
1 Platoon Commander Lt. Braun		(211)
2 Platoon Commander Lt. Feineisen		(221)
3 Platoon Commander Lt. Zwack		(231)
3 Company	Oblt. Franz Kretschmer	(301)
1 Platoon Commander Ofw Issler		(311)
2 Platoon Commander Ofw Schwarz		(321)
3 Platoon Commander Lt. Geoggerle		(331)
Versorgungskompanie (Maint Supply)	Hptm Helmut Ulbricht	
Werkstattkompanie (Workshop)	Oberleutnant Dipl. Ing Karl Schulte	

29.3 Jagdtiger deliveries in November 1944

Delivered to	Transported on	No.	Type	Chassis No.	Tactical No.
Ersatzheer	*8/11/44*	*1*	*(P)*	*305012* [1]	314
s.Pz.Jg.Abt.653	*18/11/44*	*1*	*(H)*	*305019*	?
s.Pz.Jg.Abt.653	*18-24/11/44*	*3*	*(H)*	*305020*	331
			(H)	*305021*	?
			(H)	*305022*	323
s.Pz.Jg.Abt.653	*24-28/11/44*	*4*	*(H)*	*305023*	113
			(H)	*305024*	134
			(H)	*305025*	?
			(H)	*305026*	?

29.4 Jagdtigers built in Nibelungen Werk in November

There were 6 Jagdtigers completed in November against the program for 20.

Chassis No	*305026*
	305027
	305028
	305029
	305030
	305031

29.5 Hulls built by Eisen Werk in November 1944

There were 2 Hulls completed in November giving a running total of 90.

There was no bombing of factories in November.

[1] Jagdtiger 305012 was later taken on by 653 making seven Porsche vehicles with the unit at this time.

30

Schwere Panzerjäger Abteilung 653 December 1944

30.1 December 1944 synopsis

1 December The 653's Jagdtigers are still at Battalion H.Q. in Döllersheim. The German Army is gathering in the Ardennes in preparation for a major offensive. The 1 & 3/653 companies are to be part of this.

5 December 16 Jagdtigers are put on 3 trains destined for this new offensive. They leave Döllersheim heavily camouflaged.

9 December The 3 trains avoid air attack by hiding in a tunnel near St. Goar.

12 December 2 trains unload at Wittlich - Engerohr, 50km behind the front. The 653 Abt are to be behind 6 SS front. The third train is still at St Goar.

16 December In foggy conditions, the German attack in the Ardennes begins.

19 December From Wittlich, 653 request 3 trains to move forward to Blankenheim. Only one train arrives, heavy bombing in Moseltal between Trier and Cochem stops the other two. Six Jagdtigers are loaded onto the single train.

21 December The train unloads at Blankenheim. 653 are to deploy at Gemund, Kall and Schleiden. Command awaits orders to deploy.

23 December Fresh orders are received for the transfer of the 6 Jagdtigers from Blankenheim to Zweibrücken for OPERATION NORDWIND.

28 December The Ardennes offensive runs out of steam.

29 December Train sent from Blankenheim to take the 6 Jagdtigers to Zweibrücken. Twelve Jagdtigers are ordered to leave Wittlich on road march for Boppard. This turns out to be a disastrous journey.

30 December An order deploys 653 to the 17th SS Panzer Grenadier Division for OPERATION NORDWIND.

31 December A train with 2 Jagdtigers from 3.653 arrives at Zweibrücken. A train with 3 more Jagdtigers leaves Döllersheim for Zweibrücken.

30.2 Jagdtigers used in the Ardennes Offensive - fact or myth?

There have been many stories of Jagdtigers having been extensively used in the 'Wacht Am Rhein'. Up to now none can be substantiated by photographs. Here I shall explore some of the stories and explain the facts.

Hitler's Plan

America tries to become England's heir; Russia tries to gain the Balkans. Even now these states are at loggerheads. If now we can deliver a few more blows, then at any moment this artificially bolstered common front may suddenly collapse with a gigantic clap of thunder. It is essential to deprive the enemy of his belief that victory is certain.

Adolf Hitler
12 December 1944

'Wacht Am Rhein' was Hitler's last big gamble. It was to be an offensive launched against the Allies in the west. The purpose was to reach the coast at Antwerp by driving a wedge between British forces in the north and the Americans in the south.

His plan was to break through in the Ardennes with a huge Panzer force and push north and west up towards Antwerp. In the preliminary planning phase, concern had been shown by Hitler and his Field Commander, Jodl, over the strong concentration of troops and equipment in the Aachen area and along the entire northern flank of his intended break-through point. It was obvious that there would be no point in pushing out to the coast if the flanks could not be protected. If this were the case, the whole plan was likely to fail with the loss of the valuable armored elements.

In order to secure the northern flank, Hitler decided to use the schwere Panzer Jäger Abteilung 653 with their new tank de-

Plate 233. Two Jagdpanzer IV Vomag's knocked out near Chenogne in the Bastogne area in December 1944. These vehicles were sometimes confused with Jagdtigers (Bastogne Historical Research Center).

(Courtesy Jeanco via Jean-Paul Pallud)

stroying Jagdtigers. The plan was for the Jagdtigers to deploy north of the penetration-area. They were link up with paratroops that were to be dropped behind enemy lines. Thus forming a defensive hard shoulder running generally east from a point south of Liege through Verviers to Monschau to protect the northern flank of the 6th SS Panzer Armee. It was a plan designed to hold off any Allied counterattacks attempting to reinforce the battle area from the north. The plan did not work. Von der Heydte the paratroops commander was captured as he and some of his men marched into Monschau to link up with the Jagdtiger Battalion. The town was still in American hands.

Allied reinforcements from the north were quickly reaching the battle area. When reports of this were given to Hitler, he promptly asked about the Jagdtigers. Herbert Buechs, at OKW, gave the news to the Fuhrer. "A check has been made," he said. "The transport trains bringing the Jagdtigers forward from Döllersheim have been stopped by air attacks on the planned rail route."

The transport trains of 653 went via Ludwigshafen/Mannheim across the Rhine. One train got as far as the Eiffel area, the Jagdtigers were unloaded and driven into the woods, but they did not receive any further orders to go into action in the Ardennes. However, some of the supply vehicles commanded by Oblt Ulbricht did get to the planned forward assembly area.

Because of this delay, Hitler who was keen to use the new weapon decided that the Jagdtigers transferred south to be used in 'OPERATION NORDWIND'. This was to be a classic pincer movement to recapture Strasbourg due to start on 1 January 1945.

30.3 Reports of Jagdtigers in the Ardennes

I have traced several Allied accounts alleging the use of Jagdtigers in the Ardennes Offensive.

Report 1 A Jagdtiger was reported being used in the village of Bures on 3 & 4 January 1945, supposedly assigned to the Panzer Lehr Division. This was undoubtedly a Jagdpanzer IV (Vomag) which was assigned to Panzer Jäger Abteilung 130. An earlier plan, in March 1944, to equip this unit with Jagdtigers was abandoned.

Report 2 A Jagdtiger was supposedly knocked out in Maldingen, west of St Vith. The only evidence for this is the testimony of a local resident, Mr. Jeanco who, at the request of Jean Paul Pallud, in 1988, drew from memory the vehicle he recalls seeing. Mr. Jeanco is confident about the accuracy of his recollection but this evidence is at best slender.

Report 3 A knocked out Jagdtiger reported to have been seen in the Ardennes, by Colonel-George Forty in 1948. This vehicle cannot be substantiated by photographic evidence. Neither can Colonel-Forty remember the exact location. He may have been mistaken about the location. The vehicle he recalls could have been one used in the Alsace.

Conclusion

The evidence is conclusive. According to the testimony of surviving members of s.Pz.Jg.Abt. 653, Jagdtigers were *not* used in the 'Wacht Am Rhein'.

Such an unusual armored vehicle would have created much interest among the Allied Intelligence staff. Photographs would have been taken with an on-site analysis being made. Bearing in mind, that at this stage of the war, Jagdtiger had not been encountered by the Allied troops who only knew of its existence through drawings that had been circulated to the front line troops by British Intelligence MI 10 in late 1944! If these drawings are compared to the Panzer IV (Vomag), it is easy to see how the confusion could have occurred. At the time, the Jagdtiger's existence was known to the Allied forces and a high priority was assigned to recovering an example.

Recorded evidence states: Only one Jagdtiger had been lost in action prior to 15 March 1945. This was at Rimling (Alsace) on 9 January 1945, an area that remained in German hands until late February. Additionally all Jagdtigers issued for training and testing during this period are accounted for.

30.4 Report 25 December 1944

Telegram: To OKH/Gen. Insp. d. Pz. Truppen
From, A.O.K.L.
Subject s.Pz.Jg.Abt. 653
Special tracks for the Porsche suspension (Abt. Ordered) still with seven *Jagdtigers Porsche, the rest are with Henschel suspension.*

30.5 Major transport problems throughout December 1944

For the build-up of the great Offensive on the Western Front, 16 Jagdtigers were put on 3 trains at Döllersheim on 5 December. The vehicles were heavily camouflaged with coniferous foliage. Because of heavy bombing by the unopposed Allied Air forces, the trains moved only at night and were hidden in tunnels during the day. Of these 3 trains, only 2 reached the first assembly area at Wittlich, the other was stopped in Moseltal between Trier and Cochem because of bomb damage to the lines. A further train, on the 19 December, loaded with 6 Jagdtigers from Wittlich and took them to Blankenheim. This transport fiasco prevented Jagdtigers being used in the Ardennes Offensive.

On 23 December, OKH issued fresh orders transferring s.Pz.Jg.Abt.653 to the Zweibrücken area ready to take part in OPERATION NORDWIND. A train was arranged and took the 6 Jagdtigers from Blankenheim to Zweibrücken. 2 of the 6 had already broken down with major mechanical problems during loading and unloading.

Plate 234. Porsche Jagdtigers of the 1/s.Pz.Jg.Abt 653 on rail transport for the Ardennes offensive, 5 December 1944 (Karlheinz Münch).

Plate 235. Jagdtiger of the 3/653 in Blankenheim/Eiffel for the start of the Ardennes offensive – this was as far as they got (Karlheinz Münch).

On 29 December, because of non-availability of railway flatcars OKH order the other 12 Jagdtigers to drive the 90km from Wittlich to Boppard, by road. This was expected to be a 3-day march. These 12 Jagdtigers, some of which were Porsche suspension vehicles, never reached Boppard; they broke down through various mechanical faults. This fact later instigated an investigation by the Inspector General Pz. Tr. West who sent a team of specialists from the factories to investigate and report on the extremely high number of breakdowns. Two Jagdtigers did reached Boppard, on 1/2 January 1945.

Two of the Jagdtigers from Blankenheim had also broken down when they reached Zweibrücken, on 2 January 1945. A further 3 Jagdtigers, all operational, were dispatched from Döllersheim to Zweibrücken, on the last day of December 1944.

On 30 December 1944, an order from Hitler put s.Pz.Jg.Abt 653 with the 17th SS.Pz.Grenadier Division to start the offensive at 2300 hours on the night of 31 December 1945. Only 2 Jagdtigers were available near the assembly area at the time.

30.6 Numbering

In December/January all the Jagdtigers belonging s.Pz.Jg.Abt 653 had been camouflage-painted and tactical numbers were sprayed on all 42 Jagdtigers; cardboard masks were used by the Maintenance Company.

The numbers used were:

		111	112	113	114	115 [1]
101	102	121	122	123	124	
		131	132	133	134	
		211	212	213	214	
201	202	221	222	223	224	
		231	232	233	234	
		311	312	313	314	
301	302	321	322	323	324	
		331	332	333	334	

30.7 Jagdtiger deliveries in December 1944.

Delivered To	Date	No	Type	Chassis No	Tactical No
s.Pz.Jg.Abt.653	*7-11/12*	*3*	*(H)*	*305027*	?
			(H)	*305028*	?
			(H)	*305029*	?
s.Pz.Jg.Abt.653	*8-11/12*	*1*	*(H)*	*305030*	?
s.Pz.Jg.Abt.653	*8-12/12*	*1*	*(H)*	*305031*	?
s.Pz.Jg.Abt.653	*8-12/12*	*1*	*(H)*	*305032*	?
s.Pz.Jg.Abt.653	*29/12/94*	*1*	*(H)*	*305033*	?

30.8 Jagdtiger built by Nibelungen Werk in December 1944

There were 20 Jagdtigers completed in December against a program of 20.

Chassis No:	*305032*	*305042*
	305033	*305043*
	305034	*305044*
	305035	*305045*
	305036	*305046*
	305037	*305047*
	305038	*305048*
	305039	*305049*
	305040	*305050*
	305041	*305051*

Modifications: Addition of extra space track-hooks to hull sides in December.

30.9 Hulls built at Eisen Werk in December 1944

There were 8 Hulls completed for the month giving a running total 98.

There was no Bombing in December on the factories.

[1]Karlheinz Münch has a photo showing a Jagdtiger at Schwetzingen carrying the tactical number **115**; this could have been applied by mistake or it was a replacement for the Porsche Jagdtiger removed from their inventory and allocated a new number or even a consolidation of the 1/653?
Note: By the end of December, 653 have received 27 Jagdtigers and the 1/653 and 3/653 companies are almost combat ready.

Plate 236. The production line, in December 1944. These Jagdtigers nearing completion are possibly for 2/s.Pz.Jg.Abt 653 (Walter J. Spielberger).

31

Schwere Panzerjäger Abteilung 653 January 1945

31.1 January 1945 synopsis

1 January Coincidental with the start of the New Year, Hitler launched his second Major offensive, "OPERATION NORDWIND", and an attempt to recapture Strasbourg. Part of 1/653 in rail transit to Zweibrücken, rest of 1/653 is in the Mosel Valley.

2 January 3/653 starts to assemble at Mittelbach near Zweibrücken, with 2 Jagdtigers.

4 January Part of 1/653 is at Seyweiler their first 4 Jagdtigers arrive by train.

9 January First attack by 1/653, with only 3 operational Jagdtigers they are fighting alongside 17th Panzer Grenadier Division in the area of Rimling. A Bazooka hit destroys one Jagdtiger No 134 when making an assault, all 6 crew are killed.

10 January Part of 1/653 at Neideraubach near Zweibrücken. A team of specialists sent to investigate the high number of Jagdtiger breakdowns in the Mosel valley. Hitler orders second Jagdtiger attack.

11 January Unit near Altenwald.

13 January The 2/653 take on their last Jagdtiger. Equipping of 653 is complete with their heavy weapons (42 assigned). First field inspection of broken down vehicles.

14 January Second field inspection of failed vehicles.

16 January Inspection report sent to OKH on Jagdtiger problems.

17 January Part of the 3/653's Jagdtigers attack bunkers at Auenheim.

20 January The 2/653 in arrive by rail in the Buhl area.

21 January The 2/653 is in Neustz by Buhl.

22 January Transport arrangements for 653 report to OKH.

23 January Part of 2/653 cross the Rhine near Drusenheimer on ferries.

25 January Hitler calls off NORDWIND: line stabilizes along the Moder River.

28 January Part of 3/653 still in Auenheim - Roschwoog area.

29 January The elements of 2/653 ferry back across the Rhine to Sessenheim - Buhl.

30 January The 2/653 is in Kappelwindeck near Buhl.

31.2. OPERATION NORDWIND - Himmler's offensive

"The world must know that this state, will, therefore, never capitulate!" -Spoken by Hitler on 1 January 1945.

'OPERATION NORDWIND' is often referred to as the Second Battle of the Bulge - It started on 1 January 1945.

Eight German Divisions attacked against the thinly held U.S. Front, in Alsace Lorraine, the northern prong of a pincer movement to recapture Strasbourg. The southern prong of the pincer was launched across the Rhine near Obenheim.

The 1& 3/schwere Panzer Jäger Abteilung 653, with 6 operational Jagdtigers was assigned to 17th Panzer Grenadier Division and acted as reserve for the northern attack.

Because of the precarious position of the U.S. 7th Army, north of the Moder River, Eisenhower gave the order to retreat and ordered that Strasbourg should be evacuated.

This provoked an angry reaction from General de Gaulle, who said that he would occupy it and take over the defense. Eisenhower threatened to cut off supplies and ammunition to the French Forces.

The German attack lasted 20-days. It halted at the Moder River after a bitter fighting through the snow-covered hills of the Forest of Haguenau.

96. Anl. z. Generalkommando XIII SS A.K. Ia 8/45 gKdos v. 6.1.45 s. Pz. Jg. Abt. 653

Umlauf 5 Verband: ~~17. SS Pz. Gren. Div. „G. v. B."~~

Meldung vom: 4.1.1945 ~~1944~~ Unterstellungsverhältnis: 17. SS „G.v.B."

Zu Ia 58/45 geh.

1. Personelle Lage am Stichtag der Meldung

a) Personal:

	Soll	Fehl	
Führer	28 u. 1 Bea.	3	xx)
U-Führer	261 u. 4 Bea.	-	
Mannsch.	618	-	
Hiwi	13	-	
Insgesamt	925	3	xx)

c) in der Berichtszeit eingetroffener Ersatz:

	Ersatz	Genesene
Führer	-	-
U-Führer u. Mannsch.	31	4

b) Verluste und sonstige Abgänge
in der Berichtszeit vom 1.12.44 bis 31.12.44

	tot	verw.	verm.	krank	sonst.
Führer	-	-	-	-	-
U-Führer u. Mannsch.	2	5	-	4	6
Insgesamt	2	5	-	4	6

d) über 1 Jahr nicht beurlaubt:

Insgesamt	31 Köpfe 3,3 % d. Iststärke		
davon	12—18 Monate	19—24 Monate	über 24 Monate
	31	-	-
Platzkarten im Berichtsmonat zugewiesen	-		

2. Materielle Lage:

		Gepanzerte Fahrzeuge							Kraftfahrzeuge				
									Kräder			Pkw	
		Stu. Gesch.	III	IV	V (Jagd-P.)	VI (Jagd-tiger)	Schtz.Pz. Pz. Sp. Artl. Pz. B. (o.Pz.Fu Wg.)	Pak SF	Ketten	m. angeh. Bwg.	sonst.	gel.	o
Soll (Zahlen)		-	-	-	5	45	10	-	14	-	6	37	2
Einsatzbereit	zahlenm.	-	-	-	-)	6 =)	-	-	-	-	3=)	=)15	1
	in % des Solls	-	-	-	-	13,3	-	-	-	-	50	40	50
in kurzfristiger Instandsetzung (bis 3 Wochen)	zahlenm.	-	-	-	-	1	-	-	-	-	1	10	2
	in % des Solls	-	-	-	-	2,2	-	-	-	-	16,6	26	100

		noch Kraftfahrzeuge							Waffen			
		Lkw				Ketten-Fahrzeug						
		Maultiere	gel.		Tonnage	Zgkw. *)	Zgkw. **)	RSO	s. Pak	Art-Gesch.	MG ( )	sonstige Waffen (2 cm Fla Vierl.)
Soll (Zahlen)		6	67	47		8	13	-	45	-	68(59)	3
Einsatzbereit	zahlenm.	6	7=)	22 =)		2=)	2=)	-	6	-	16(6)	3
	in % des Solls	100	10	48		25	15	-	13,3	-	23(10)	100
in kurzfristiger Instandsetzung (bis 3 Wochen)	zahlenm.	-	2			-	-	-	1	-	-	-
	in % des Solls	-	3			-	-	-	2	-	-	-

*) Zgkw. mit 1—5 t **) Zgkw. mit 5—15 t
( ) davon MG 42

3. Pferdelage: Soll: ---
Ist: --- davon Panjepferde ---

Anmerkungen siehe Rückseite!

Anl. zu Nr. 117 / 45 geh.
Gen. Insp. d. Pz. Tr.

The 653 were to assemble in the northeast perimeter of the Haguenau Forest over the next 2 months. Hidden in the forest the unit waited to counter the anticipated U.S. counter-attack. While hidden and heavily camouflaged, the unit was still attacked by Allied fighter bombers.

On 25 January, Hitler ordered that the attack against the Lower Vosges and Lower Alsace be suspended, because the forces being used to attack the Moder line were required as reserves behind future defense efforts. This included 653, which was starting to arrive in the Landau area.

31.3 Report Dated 4/1/1945 s.Pz.Jg.Abt.653 (17.SS Gr.B)

1. Personnel Log a. Total 925 men, 28 officers, 261 NCO, 618 men.

1/12/44 to 31/12/44 b. Casualties and departure, 2 dead, 5 wounded, 4 ill, 6 others.

c. In the promised reserves list, 31 NCO's, 3 convalescence.

2. Material Log Jagdtiger (quota 45, operational 6, short-term repair 1).

Berge Panther (quota 5, operational 0, repair 0).

Flak Pz 1V (not indicated).

3. Notes: x, Compilation of quota and deficit figures in accordance with order OKH Oh.H.Rust. u.BdB / AHA / In 6 [VIII E] Nr.17654 / 44 secret 28 . 11. 44. Still awaiting arrival of personnel or delivery of vehicles for flack platoon, these not included.

xx, One officer position through FhJ. Ofw engaged and 2 officers are for the Battalion Commander.

=, Further Kfz coming forward on rail transport will be assigned to supply unit. Half-tracks for Battalion will be replaced by wheeled vehicles (le PKW bzd).

4. Short assessment by the Battalion Commander

***Training:** 80% of the men originate from the original Battalion, the practicality and value of the theoretical and practical training will become known in operation. Changes including the 20% replacements, their training on troop practice ground, along with technical and tactical aspects has ended. At present, experience with the Panzer is not complete, as they are all not available. Schooling in shooting and fighting practice, including the main armament has been completed the results are good.*

***Morale of the troops:** The troops are above average with their willingness and cheerful endeavor to bring successful results with these new weapons. After almost 4 weeks in transportation, with constant postponements due to difficulties from employing temporary means, this has adversely affected our strength against attack.*

***Special difficulties:** Breakdown of transport system, abnormal land marches to mission area (over 130 km). Only a small number got to operation area as soon as possible, but still missing out on K st N, I absolutely compelled auxiliary means, but this still inhibited the first mission and was the cause for present lack of success. The principles of battle and operation because of the above have almost without exception become disregarded. The decentralizing in the area of 2 Army Group with a scattering of the unit has weakened our operational readiness and overtaxed our strength. This has been caused by transportation from three different railheads.*

***Judgment:** The complete Battalion is corresponding to your personnel and material situation for each appropriate attack and defense duty*

Major and Battalion Commander.

31.4. The first Jagdtigers in combat

Six Jagdtigers from 1 & 3/653 had been transferred to the Zweibrücken area to take part in OPERATION NORDWIND. They were placed with the 17th Panzer Grenadier Division, which was commanded by SS-Stan Fhr Linger. Their section of attack was to be through Rohrbach and southeast towards Sarrebourg.

The first 2 of the 6 Jagdtigers were from 3/653 and they arrived in arrived Zweibrücken on 31 December 1944.

On the 4 January, the 1/653 with 4 Jagdtigers unload near Zweibrücken, they set off in blizzards from Mittlebach along roads towards Seyweiler. At this stage the 17th Panzer Grenadiers were using their Stug III's up front, fighting and progress was slow mainly due to bad weather.

By 5 January, the Jagdtigers were in Seyweiler, still in reserve. One of the 1/653 Jagdtigers was having mechanical difficulties and dropped out. It was later towed to Rohrbach railway station for recovery. A further 3 Jagdtigers had arrived at Zweibrücken from Döllersheim, these also belonged to 3/653.

By Tuesday 9 January, the 3 Jagdtigers had only had the opportunity to destroy a single Sherman. Heavy snow was everywhere. The 3 Jagdtigers were attacking across snow covered fields towards Rimling when the right hand tank, No 134, commanded by Uffz Fritz Jaskiela exploded. U.S. infantry in snow trenches had hit it with a Bazooka. The Jagdtiger was hit in its right side by a Bazooka, which exploded its ammunition. It was totally destroyed with the loss of all 6 crew. This was the first Jagdtiger lost in combat (chassis No 305024).The front at the time was in total confusion.

Plate 237. Heavily camouflaged Jagdtiger near Zweibrücken, in early January 1945 (Karlheinz Münch).

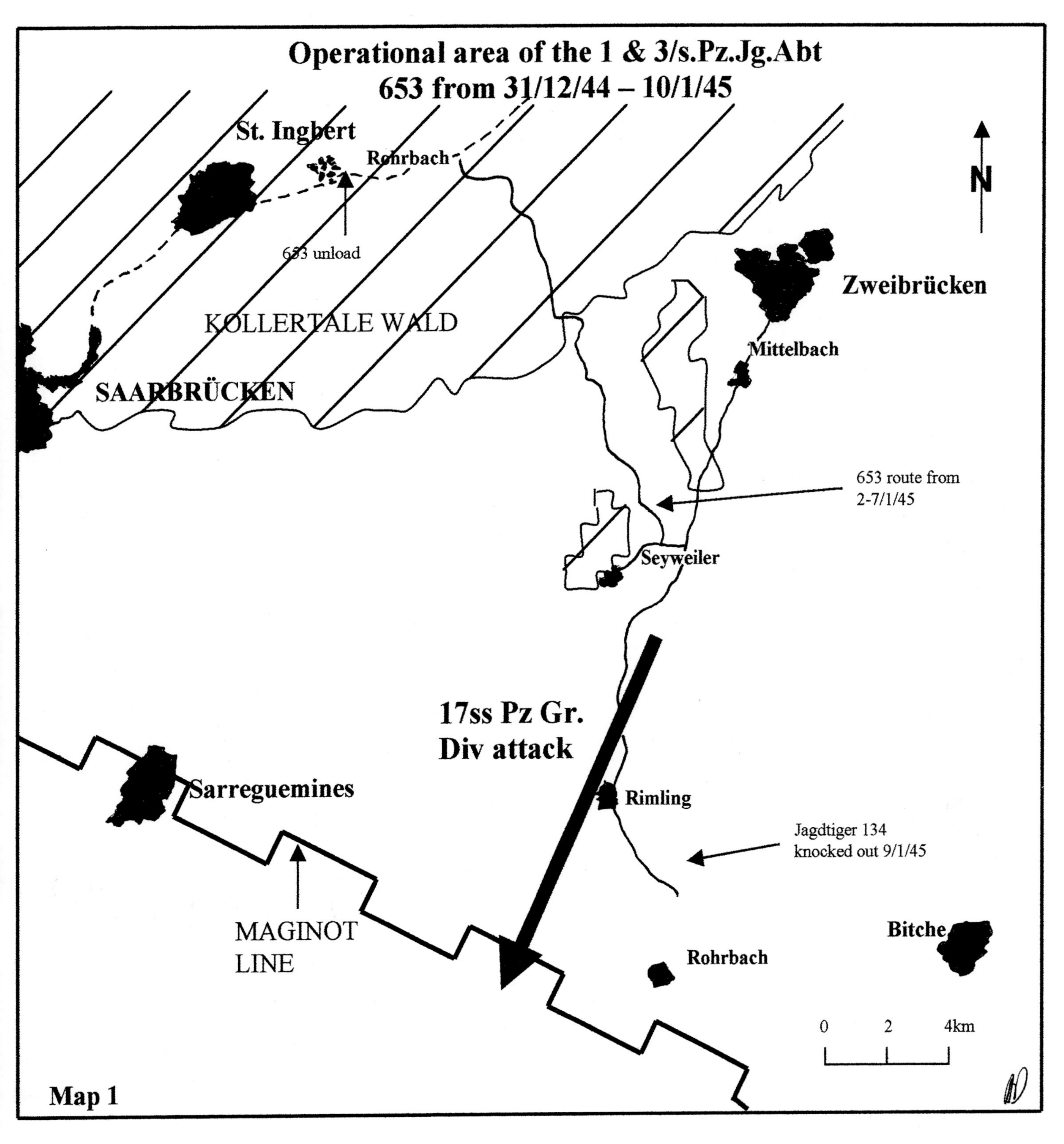

The commander of the second Jagdtiger, Fw Reinhold Schlabs, 250 meters to their left, spun his Jagdtiger No 123 and hit the infantry with machine gun fire and explosive, (percussion mortars), which along with the accompanying Panzer Grenadiers, dealt with the U.S. troops. The remaining 2 Jagdtigers saw very little action and were eventually loaded on a train, on 5 February, at Rohrbach for transferal to Bellheim.

The company commander, Lt. Haberland, was traveling to the Mosel Valley at this time and was not directly involved with the attack at Rimmling.

The Jagdtiger was not reported as destroyed, until 16 January, when the reason for its destruction was stated as "Not known".

The remains of the Jagdtiger were overrun by the U.S. Forces, in mid February 1945. It was extensively photographed, on 28

Plate 238 - 244. Photo sequence of Jagdtiger tactical number 134, chassis number 305024, after it was knocked out. Further demolition was effected to remove its potential use to the Allied intelligence service. This was their first examination of a Jagdtiger (U.S. Army).

February; the first Jagdtiger inspected and photographed by Allied Forces.

A photo caption and an American record, claimed the Jagdtiger as being destroyed by an M36 tank destroyer, according to crew sources this is not true!

The same day as the first Jagdtiger loss, as mentioned earlier, the front was in utter confusion. 17th Panzer Grenadier Division had had to withdraw from breakthrough positions. Their commander, SS - Standartenfuher Hans Linger, was taken prisoner.

The Jagdtiger objective in the attack had been to break any strongly defended positions or bunkers of the Maginot line in the area of Rohrbach-Les-Bitche.

In the event these bunkers were overrun by an infantry attack which had made use of the blizzards. The Jagdtigers were not now required for the assault on the static fortifications in this sector.

All the Jagdtigers in the Zweibrücken area were later sent by rail to the Landau area, in early February.

After No 134 had been knocked out, the troops from 653 buried their dead; the medical staff treated the wounded. The burned out Jagdtiger which was now, of no possible use to the Battalion, was blown up to further restrict its future value to the Allied intelligence staff.

Report 2 March 1945 US 12th Army group reports on first captured Jagdtiger (305024), the following photos were taken.

31.5 Operational Report - 9 January 1945

On 9 January 1945, the same day that 1/653 went into action for the first time, a report was made to the Inspector General of the Tank Troop West, Reference s.Pz.Jg.Abt. 653 Jagdtigers:

General

O. U. the 9.1.1945

the Panzertruppen West

-Abt. Ia Nr. 371 / secret

Reference: *s.Pz.Jg.Abt.653 (Jagdtiger).*

To

Ob.-West

Gen. Insp.d.Pz.Tr.

Two Jagdtigers stood ready for action in the area of Boppard and ordered to be loaded onto rail transport the same evening 9.1.45.

Two Jagdtigers were in the area of Emmelshausen, 12km west of St.Goar. Both had engine trouble because of low oil-pressure.

There were two Jagdtigers, in Godenroth, 15km south west of St.Goar, one with steering damage and the other with cooling-fan damage.

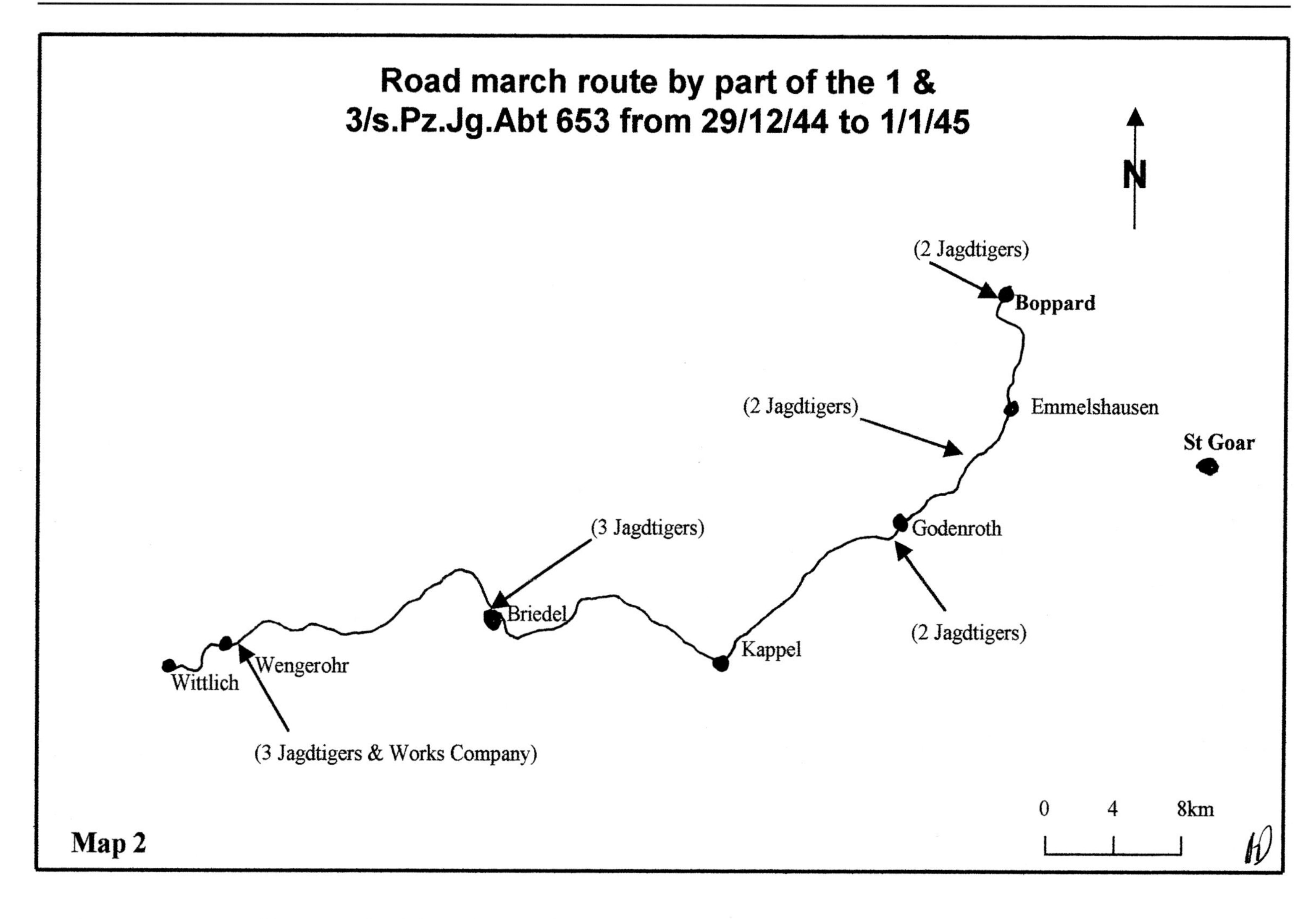

There were 3 Jagdtigers, in Briedel on the Mosel, one had engine damage due to low oil pressure, one had an electrical fault and loss of cooling-liquid, the third had a damaged valve and piston on the engine.

There were 3 Jagdtigers in Wengerohr. One had transmission damage, the other two had low oil-pressure.

At this time, the 1/2 Workshop Company of 653 was working on these last 3 Jagdtigers at Wengerohr. The work was difficult because they had no cranes and had to rely on the use of the train mounted railway crane in Wengerohr. They expected 2 Jagdtigers to be repaired and arrive in Boppard by the evening of 12 January 1945.

For General of the Panzertruppen West
The 1. General staff officer
I.V.

Major

Plate 245. Jagdtiger No 314 (Porsche) in the Mosel area waiting for repair work (Karlheinz Münch).

On receiving this report, Major General Thomale ordered an immediate investigation into the high number of reported mechanical breakdowns.

31.6 Hitler orders Jagdtiger attack

On 10 January orders were issued by the Fuhrer:

(20.45 H.G.H ObW) that two Jagdtigers on their way to s.Pz.Jg.Abt. 653 were to be diverted immediately to the 10th SS Pz. Div to attack bunkers.

Within half an hour a response was sent to 653:

H.GR.G 10 January 1945 21:15 H.GR. stated XXXIX Pz.K. and Kdr. of the s.Pz.Jg.Abt 653 should have two Jagdtigers conveyed to the Panzer unit. Both Jagdtigers were to be conveyed

Plate 246. A typical Maginot lines bunker near Kauffenheim, on January 6 1945 (U.S. Army).

as quickly as possible to the area south west of Lauterburg and have ammunition ready to go straight into action.

On 13 January 1945 the remaining vehicles of s.Pz.Jg.Abt. (Jagdtiger) 653 (3.k.p) were immediately to be acquired by OB. West from Döllersheim via Pz Base middle (Buhl area). By requisition of E-transport, they were to be transported. This order OKH/ Gen S td H/Op Abt (roem 3), 650/45 secret had been passed on through Major Wolfram i.G.u.stellv.Gr.Ltr.

The first train with five Jagdtigers left Buhl on the 14 January; the Rhine was crossed at Germesheim.[1] The Jagdtigers were then unloaded at Lauterbourg on the night 15/16 January. This was followed by a 25km-road march through Mothern and Seltz to Roeschwoog; one Jagdtiger broke down on route.

The five Jagdtigers were assigned to XIV SS troops and were available and ready in the area of action by the morning of 17 January. They were used to support an infantry attack against part of the Maginot bunker line near Auenheim.

Part of the 22 January report stated:

On 18 January four bunkers were attacked at a range of 1000m. The firing proved to be extremely accurate with one bunker turret burned out after two shots and similar hits were made at the bunker firing slits. One Allied Sherman fired back and was set on fire.

In total, 46 rounds of Spr.GR. and 10 rounds of Pz.GR. were used. There were no casualties among the Jagdtiger personnel.

The five Jagdtigers remained with XXXIX Pz. Korps for the duration of OPERATION NORDWIND in the area of Drusenheim; they were not able to get across the Moder River with the German forces in the Bridgehead around Bischwiller.

31.7 Specialist report on investigation of major mechanical failures

Oberstlt Johannis, the Motor Vehicle Officer, reported the following on 16 January 1945 to the Inspector General of the Panzer Troop:

H. Qu. OKH the 16.1.45.
Bb. Nr. 1. 290/45 secret
Journey report

Ref: Jagdtigers of the s.Pz.Jg.Abt. 653

On the orders of General Thomale, on 13 & 14 January 1945, 2 specialists from Henschel and Maybach and myself visited broken-down Jagdtigers of the s.Pz.Jg.Abt. 653, to investigate and identify the damage.

Findings

10 out of the 16 in the area of Wittlich (Mosel) were to be dispatched by road march, but only got to Boppard on the Rhine (90km drive by road) - they broke-down along the route.

6 with heavy damage

Four Jagdtigers (No's 305010, 305014, 305017 and 305031) with steering unit bearing failure. Two Jagdtigers (No's 305012 and 305025) with gearbox damage.

4 with minor damage

No 305011 had electrical short circuit resulting in engine fires.
No 305022 had damage to the valve-timing gear.
No 305009 with cooling-fan shaft breakage.
No 305019 with loss of gearbox oil due to paper gaskets.

Repairs

The repair parts were ordered immediately. Gen.d.Pz.Tr. West sent a crane wagon to assist with the repairs, together with 3 tank mechanics, and 1 factory specialist was also sent to the unit.

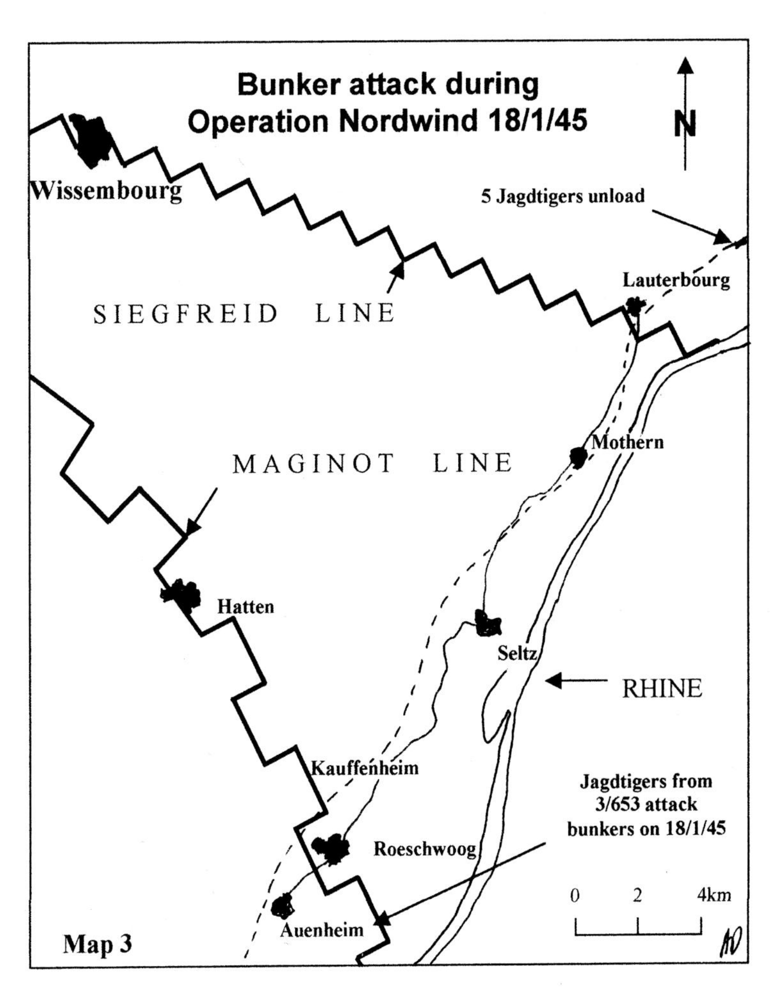

The 4 lightly damaged vehicles could be repaired in 1 to 2 days on arrival of the parts. The 6 others, heavily damaged, would take a further 1 to 2 weeks.

Cause of Breakdowns

Because the Jagdtiger is 10 tons heavier than the Tiger II, it is much more sensitive and therefore susceptible to overloading.

The training of the drivers and technical personnel is not sufficient even though they have worked in the Nibelungen Werk - they are not familiar with the Henschel motor or with the Olvar drive and therefore cannot cope with the problems.

Therefore recommendation:

That a specialist be sent in immediately from the Henschel Firm that understands the whole of the motor drive system to teach the drivers and technicians of the s.Pz.Jg.Abt. 653. This should take approx. 2 weeks to teach them.

The personnel from Henschel should also teach the drivers from the Jagdtigers in the Kassel area.

The s.Pz.Jg.Abt.653 did not have the specialist tools and therefore in most cases could not do the work themselves.

On 6 of the vehicles they had to investigate loss of tools.

There was no incentive not only with the drivers, but also with their superiors (Oblt. Haberland, Chef I. Kp./653 and Baurat Jörger, who was acting Unit Engineer). I urged them to, as quick as possible, make further preparations to their vehicles, or at least get them ready for the necessary repairs before hand. Most of the vehicles stood for 5 days without any preparation or inspection to find out/assess the damage.

This impossible position was brought to the attention of the person in command and also to Gen.d.Pz.Tr West who was asked to keep a check on the unit.

The fighting unit was divided between the Wittlich and Zweibrücken area, together with limited equipment available to the works unit and only having one crane, it was impossible to supply, maintain and repair all the Jagdtigers.

At Abt.653 there were only Jagdtigers works units in the Zweibrücken area, there were no Jagdtiger work units in the Wittlich area. The fighting troops were not permitted to load the equipment onto rail transport at any time.

General Opinion Statement

The Kp. Chef. Abt, Ing. Schirrmeister and drivers from 1.Kp. and 3.Kp./653 had marching performance of 30-40km a day, which was reasonable. The route from Wengerohr to Boppard (approximately 90km) was expected to be a 3-day march. Some of the Jagdtigers completed this distance in 2 days.

This was put down to the training in the Nibelungen Werk, whose aim was to take special care of the equipment. Some 653 personnel did not have confidence in the Jagdtiger's ability to cover such a distance.

Back at the Henschel factory it was pointed out through Gen.d.Dr. Ing Stieler Von Heydekampf, Dir Pertus and others, that the Jagdtiger was the best vehicle on the front, and was having good results, therefore they demanded to build more than the planned run of 150 Jagdtigers.

I explained to the Inspector General of the Tank Troops that we needed good verification of the Jagdtigers achievements, as up to now very little is known. The following statement from Oblt Haberland, Kp.Chef 1 /653, "up to now there were only a few Jagdtigers in action in the Zweibrücken area. There was only 1 Sherman destroyed, 1 Jagdtiger was totally lost through explosion, reason not known".

Johannis

31.8 Transport arrangements of the s.Pz.Jg.Abt. 653 on 22 January 1945

From (Brückenkopf) the assembly area of the XXXIX.Pz. Korps:
5 Jagdtigers - 4 ready for action, 1 on long-term repair

In the area of Buhl:
8 Jagdtigers - 4 ready for action, 2 on short-term repair
2 on long-term repair

In transport from Saarpfalz:
5 Jagdtigers

Zweibrücken area:
4 have not been transported - 3 ready for action
1 under short-term repair

Mosel/Rhein:
10 Jagdtigers - All need short-term repair

Buhl Area:	-	8 Flak-Panzer IV
Zweibrücken Area:	-	3 Bergepanzer V
Mosel Area:	-	2 Bergepanzer V

Special difficulties: Supply and repair rate. A crane, as well as repair facilities are still missing from the operation area. One entire workshop train is still employed in the Mosel. Still missing spare parts especially supplies for the repairs to transmissions. The command vehicles 9-(S.P.W) [2] are still missing, we have been expecting supply from Spandau since 5/12/1944, still not traced. Ammunition supplies still in the ammunition store. All Kfz still not in transit: 7 Sd.Kfz.10 [3], 2 Le.Lkw [4], 3 m. Lkw [5], 1 Berge Pz [6], 1 Sd. Kfz 200 [7], 22 S. Lkw [8], 1 Kfz 42 [9], 2 Sd Kfz 9/1 [10]. Also Berge Pz V with cable winch is inadequate, on main roads Jagdtigers are towed away with one Berge Pz V and two 18-ton Zg Kw [11].

At this stage the unit had an assembly area near Landau. From here 653 would become the reserve unit behind the forces in Alsace area. The whole of the Workshop Company at this time was still in the Zweibrücken area. They were however, also hindered due to the transport bringing the spare parts forward being delayed through various reasons.

Between 22 January 1945 and 5 February 1945, the train (from Saarpfalz) arrived in the Landau area. Five Jagdtigers were unloaded, it was immediately sent back to collect three Jagdtigers from the Zweibrücken area. The two Jagdtigers were collected by rail from Boppard. One Jagdtiger was driven from Lauterbourg area to Bellheim with a further four in rail transit from Roeschwoog station.

Anlage zu Reisebericht
Obstlt. Johannis

Betr.: Schadhafte Jagdtiger der s.Pz.Jäger-Abt. 653.

Bb.Nr. 1290/45 geh.

256

A. Schwere Schäden.

Lfd. Nr.	Fahrg.-Nr.	km Stand	Motor-Nummer	Schaden	Maßnahmen zur Instandsetzung
1	305 010	207	a a m 832 1373 (Auto-Union) mit 8. Kurbelwellenlager.	Kein Öldruck, Späne im Ölfilter. Pleuellager 6 und 12 ausgelaufen.	Austauschmotor bereits angeliefert.
2	305 014	174	228 a a m 832 1728 (Auto-Union)	Kein Öldruck, Späne im Ölfilter. Pleullager ausgelaufen.	Austauschmotor schon angeliefert.
3	305 031	210	229 e r e 61 352 (Maybach?)	Kein Öldruck, Späne im Ölfilter. Pleuellager ausgelaufen.	Austauschmotor im Zulauf.
4	305 017	215	a a m 832 1774 (Auto-Union)	Kein Öldruck, Späne im Ölfilter. Pleuellager ausgelaufen.	Austauschmotor ist bestellt worden.
5	305 012	ca.250	./.	Bruch derPlanetenträgerwelle im Lenkgetriebe.	Getriebe ist bestellt.
6	305 025	223	./.	Getriebe schaltet nicht. Stahlbruchspäne im Ölfilter des Getriebes.	Austauschgetriebe bestell

B. Leichte Schäden.

Lfd. Nr.	Fahrg.-Nr.	km Stand	Motor-Nummer	Schaden	Maßnahmen zur Instandsetzung
7	305 022	175	p y e 832 2585	Klopen in Zyl.-Reihe 7 - 12, vermutl.Schwinghebelwelle beschädigt.	Ersatzteile werden einem der Schadmotor lfd.Nr. 1 - 4 entnommen.
8	305 011	402 (Fahrschulwagen)	./.	Brand im Motorenraum. Elektrische Kurzschlüsse. El. Anlasser versagt.	Ersatzteile v handen. Zufü rung v. Handw u.Kranwagen veranlasst.
9	305 009	329	./.	Bruch des Kardangelenks der linken Lüfterwelle.	Ersatzteil i Zulauf.
10	305 019	406	./.	Getriebe u.Lenkung arbeiten nicht.Hoher Ölverlust durch Undichtheit am Ölfilter.	Ölfilter neu abdichten. Öl nachfüllen.

31.9 Jagdtiger deliveries in January 1945.

Delivered to	Transported on	No	Type	Chassis No	Tactical No
s.Pz.Jg.Abt.653 (27/12/44)	*2-9/1/45*	*4*	*(H)*	*305034*	?
			(H)	*305035*	?
			(H)	*305036*	?
			(H)	*305037*	(214)
s.Pz.Jg.Abt.653 (13/12/44)	*03/01/45*	*1*	*(H)*	*305038*	?
s.Pz.Jg.Abt.653 (30/12/44)	*04/01/45*	*4*	*(H)*	*305039*	?
			(H)	*305040*	?
			(H)	*305041*	?
			(H)	*305042*	?
s.Pz.Jg.Abt.653 (30/12/44)	*06/01/45*	*4*	*(H)*	*305043*	?
			(H)	*305044*	?
			(H)	*305045*	?
			(H)	*305046*	?
s.Pz.Jg.Abt.653	*13/01/45*	*1*	*(H)*	*305047*	?
s.Pz.Jg.Abt.653 (04/01/45)	*13/01/45*	*1*	*(H)*	*305048*	?
	Total 42 delivered: to s.Pz.Jg.Abt.653				
Putlos	*25/01/45*	*1*	*(H)*	305049	

31.10 Jagdtigers built by Nibelungen Werk in January

There were 10 Jagdtigers completed against the program for 20.

Chassis No: *305052*, *305053*, *305054* [12], *305055*, *305056*, *305057*, *305058*, *305059*, *305060*, *305061*

All of these Jagdtigers were assigned to the new s.Pz.Jg.Abt.512. (See chapters 37 & 38).

31.11 Hulls built in Eisen Werk in January

There were 7 Hulls completed giving a running total of 105.

Notes:

(1)The Rail Bridge at Germersheim was a two track steel span structure 318m in length and built in 1875-78.
(2)Schutzenpanzerwagen - Sd Kfz 251 Half-track,
(3)Leichterzugkraftwagen 1 ton Half track,
(4)Leichtlastkraftwagen - light half-track lorry (2 ton Maultier),
(5)Mittlerlastkraftwagen - medium half-track lorry (3 ton),
(6)Berge Panther rope winch could develop a 40 ton pull,
(7)Armored scout-car,
(8)Schwerelastkraftwagen - heavy half-track lorry (4/5 ton),
(9)Nachrichtenwerkstattkraftwagen - signals workshop truck,
(10)18 ton half-track with flat bed mounting 6 ton crane with 180 degree rotation,
(11)Recovery version (called the Bull) open body, canvas hood and 40 ton winch - Panzerbergerat 18t.
(12)See plate 6, *Jagdtiger Vol. I*

32

Schwere Panzerjäger Abteilung 653 February 1945

32.1 February 1945 synopsis

1 February Strength Report to OKH: 653 have 41 Jagdtigers, 22 operational, 19 need repair. The 2/653 is at Kapplewindeck by Buhl. The unit assembly area is near Landau.

3 February The 653 start assembling near Landau.

5 February The 1/653 is at Rohrbach - 2/653 is at Herxheim near Buhl.

8 February The 653 battalion have 32 Jagdtigers in the Landau area.

16 February Order issued by H.OKW grounding all Jagdtiger units for technical reasons.

17 February The 653 ordered south to Haguenau Forest/ Wissembourg depression area.

19 February The 653 travel south through Wissembourg, Soults and unload at Surbourg.

21 February The 3/653 deploys at Hollenhof. The last week in February sees the Jagdtigers being sent by rail from Landau to northwest Haguenau forest area. The 3 companies of 653 have different assembly areas. The Workshop Company stay at Bellheim.

22 February Order: six Porsche Jagdtigers still with 653 are to transfer to s.Pz.Jg.Kp.614.

27 February Still 2 Jagdtigers at St. Ingbert needing rail transport to assembly area.

28 February There are 39 Jagdtigers in camouflage positions await expected Allied attack. They are the reserve behind the eastern Moder Front. Radio traffic is kept to a minimum.

32.2 Report Dated 5 / 2 /45 s.Pz.Jg.Abt.653 (Heers Gruppe G).

1. Personnel log: a. Total of 1000 people, 29 officers, 271 NCO's, 682 men.
1/1/45 to 31/1/45 b. Casualties and departure (11 dead, 5 wounded, 7 ill, 4 others).
c. In the promised list of replacements, 1.

2. Material log: Jagdtiger (quota 45, operational 22, in short-term repair 19)
Bergepanther (quota 5, operational 4)
Flak Pz IV (quota 8, operational 7, in short-term repair 1).

3. Comments: x On 1/10/45 met quota in accordance with OKH / Chef H prepare and BdE / AHA In 6 [VIII E] Nr 17564 / 44 secret 28/11/44. Armored flak platoon on strength with 1 officer, 10 NCO's and 64 men. Still no K St N for the provisions for Flak platoon available. With Flak platoon added, quota figures now include 1 officer, 10 NCO, 64 men.

Short judgment by the Battalion Commander

Training: *The preparation of 1 and 3 companies is all very good, both companies have been supplied with drivers from the Russian and Italian fronts, the replacement drivers are still very low on Jagdtiger training. Both companies are fully familiar with the new weapon. The second company is a new posting in transit to here since 23/1/45. The regulars are also made up from older more-experienced soldiers. The theoretical and tactical training is fully complete and good.*

Morale of the troops: *Operationally ready and willing to bring the new weapons into action. The separate transport arrangements have unfortunately pulled the Battalion apart. Leadership and maintenance education was done in parts, as this at one time was not possible for a two-month period. Covered later, the*

71

Meldung vom 5. 2. 1945

Verband: s.Pz.Jäg.Abt. 653
Unterstellungsverhältnis: He.Gruppe G

1. Personelle Lage am Stichtag der Meldung: 1. 2. 45

a) **Personal:**

	Soll	Fehl
Offiziere	29 u.1 Bea.	3
Uffz.	271 u.4 Bea.	-
Mannsch.	682	-
Hiwi	13	-
Insgesamt	1000	3

X)
XX)

b) **Verluste und sonstige Abgänge:** in der Berichtszeit vom 1.1. bis 31.1.45

	tot	verw.	verm.	krank	sonst.
Offiziere	-	-	-	1 Bea.	-
Uffz. und Mannsch.	11	5	-	6	4
Insgesamt	11	5	-	7	4

c) **in der Berichtszeit eingetroffener Ersatz:**

	Ersatz	Genesene
Offiziere	-	-
Uffz. und Mannsch.	-	1

d) **über 1 Jahr nicht beurlaubt:**

insgesamt:	65 Köpfe 6,5% d. Iststärke	
davon: 12–18 Monate	19–24 Monate	über 24 Monate
65	-	-
Platzkarten im Berichtsmonat zugewiesen: -		

2. Materielle Lage:

		Gepanzerte Fahrzeuge						Kraftfahrzeuge				
	Stu.-Gesch.	III	Pz. (Flw.)	Berge-Pz.	Jagd-Tiger	Schtz. Pz. Pz. Sp. Art. Pz. B. (o. Pz. Fu. Wg.)	Pak. SF.	Kräder: Ketten	Kräder: m. angetr. Bwg.	Kräder: sonst.	Pkw: gel.	Pkw: 0
Soll (Zahlen)	-	-	8	5	45	10	-	14=)	-	6	37	2
einsatzbereit zahlenm.	-	-	7	4	22	-	-	-	-	6	22	2
einsatzbereit in % des Solls	-	-	87	80	52	-	-	-	-	100	62	100
in kurzfristig. Instandsetzg. (bis 3 Wochen) zahlenm.	-	-	1	-	19	-	-	-	-	-	10	-
in kurzfristig. Instandsetzg. (bis 3 Wochen) in % des Solls	-	-	13	-	42	-	-	-	-	-	27	-

	noch Kraftfahrzeuge							Waffen			
	Lkw: Maultier.	Lkw: gel.	Lkw: 0	Lkw: Tonnage	Ketten-Fahrzeuge: Zgkw. *)	Ketten-Fahrzeuge: Zgkw. **)	Ketten-Fahrzeuge: RSO	s. Pak 80	3,7 Flak	MG.	sonstige Waffen Vierl.
Soll (Zahlen)	6	67	47		8	13	-	45	4	68(58)	7
einsatzbereit zahlenm.	6	6	35		4	9	-	22	4	63(12)	6
einsatzbereit in % des Solls	100	9	74		50	70	-	52	100	92(20)	86
in kurzfristig. Instandsetzg. (bis 3 Wochen) zahlenm.	-	3	24		-	-	-	19	-	-	1
in kurzfristig. Instandsetzg. (bis 3 Wochen) in % des Solls	-	5	51		-	-	-	42	-	-	14

*) Zgkw. mit 1–5 t, **) Zgkw. mit 8–18 t
( ) davon MG. 42

3. Pferdefehlstellen:

Anmerkungen siehe Rückseite.

Anl. zu Nr. 00195 /45 geh.
Gen. Insp. d. Pz. Tr.

Wehrkreisdruckerei VIII, Breslau. 2220 L

repair of the tanks carries on with only temporary means available, (all the tanks in the Mosel Valley). Since the start, the Works Company has been split up due to transport difficulties and has been in the Zweibrücken area for three weeks without a repair task. They had worked in the operation area when required with part of the Battalion at the breakthrough point. The company chief is constantly traveling between the parts, which is with disadvantageous consequences. Apart from the breakdowns, we are in a very operationally ready state of health. Can one make clear, at this stage; judgment over the whole Battalion has not yet become appropriate.

Special difficulties: *These have already split up the Battalion, from here on special measures need to be taken. Combined with this parts of the rail transport have been dispatched for over a week, there are a disproportionately high number of breakdowns for the repair crews to recover and repair.*

Missing from the operation area:
1. The whole of the Works Company inclusive and the need for towing vehicles.
2. The supply of spare parts.

Temporary means have to be used in the recovery and supply dispositions. For the first time the Jagdtiger is in operation, for us to be effective from the above findings especially the similar occurrence of damage on the steering, gearbox, side drive, and gun alignment, this has been very disadvantageous and has delayed us. This has weakened the confidence of the troops to this otherwise outstanding Tank Destroyer. The tank driver's have already become nervous to slight damage and are trying remedies with makeshift means with combinations of exchange, which is our only present option, otherwise there will be a regrettable short stand down. Further requirements are appearing as a result for this vehicle. These all became suddenly broken down after the command through H. Gr. G. that the Battalion was to complete assembly in the area south of Landau, waiting time now apparent.

Plate 247. Jagdtiger No 314 undergoing minor repairs to its track (Karlheinz Münch).

Judgment: *8 to 10 days after the complete assembly of the Battalion, we will be suitable for each attack and defense mission. It is included in the technical report to the Gen Insp Pz Tr, that the superior office encourage a put down of the steering gearing and side drive-units, as weaknesses have resulted, otherwise we can expect a disproportionately high failure rate.*

Major and Battalion Commander.

Plate 248. Jagdtigers moving through the Haguenau area (Author).

Plate 249. Porsche Jagdtiger in the snow (Karlheinz Münch).

Plate 250. Jagdtiger arrives in the Haguenau Forest (Karlheinz Münch).

32.3 Report to OKH on 5 February 1945

1. The unit had an assembly area near Landau. At this time there were 11 Jagdtigers already at the assembly area, 10 operational and 1-requiring repair.

2. The area of Moseltal, between Boppard and Briedel, 8 Jagdtigers, 1 operational, 4 in long-term repair and 3 in short-term repair.

3. One Jagdtiger ready for action in the area of St. Ingbert.

4. Between Buhl and Landau in transit to the meeting area 19 Jagdtigers - 9 ready for action, 4 long-term repair, 6 short-term repair.

5. Two Jagdtigers are being transported from Döllersheim.

On the same date it was reported in the last few days that the enemy had attacked with tank support near Drusenheimer, Jagdtigers and several Panthers stand firm.

32.4 Report to OKH on 8 February 1945

The Panzer Reserve in the HZA on 8 February 1945 was two Jagdtigers.(1)

32.5 Report to OKH on 10 February 1945

The s.Pz.Jg.Abt. 653 with thirty-two Jagdtigers. Also notification on this date that one Jagdtiger totally lost through artillery fire.(2)

17 February 1945 OKH orders 653 south to Haguenau area.

32.6 Getting ready for operation

Throughout early February, 653 were being assembled in the Landau area. 41 Jagdtigers were being transported to the area. The Supply and Workshop Companies were also there. Grillinberger had a major task in front of him:

To get them to the new area of operations.
To put Jagdtiger Battalion into an operational state.

Plate 251. Henschel Jagdtiger on rail transport still with its combat tracks – to save time at the unloading ramp (Karlheinz Münch).

Plate 252. In close-up, the chassis number 305032 can be seen (Karlheinz Münch).

A train had to be sent to St Ingbert near Saarbrücken to collect the Jagdtiger from 1/653 along with some support vehicles.

A further train was sent along the Mosel Valley, there were already 2 Bergepanthers to recover 10 Jagdtigers from the 1/653 & 3/653. These had been stranded, since early January, after a disastrous road march from Wittlich to Boppard (90km). Only 1 of these, No 305011, had been repaired, the others were taken to the area of the Works Company at Bellheim.

4 trains with 19 Jagdtigers from 3/653 and 2/653 were making the journey between Buhl and Landau using the Rail Bridge at Germersheim to cross the Rhine. They had just completed training and gunnery practice at Döllersheim, in late January. Of the 19 vehicles, 6 needed minor repairs and 4 major repairs.

The last 2 Jagdtigers of 2/653 were being transported by rail from Döllersheim both were operational.

On 16 February, Grillinberger received the order from the Gen.Insp.Pz.Tr. that there was a patent defect in the Jagdtiger L801 steering-units. These had to be replaced, with the Jagdtigers being pulled out one at a time to fit the new steering-units which still had to be sent to the Battalion, (see following).

On 17 of February, 653 were ordered to transfer south to take up positions in the Forest of Haguenau and southwest of Soultz. Trains were used to convey the Battalion south mainly during hours of darkness on the route Landau, Wissembourg to Soultz.

By the end of February, 39 Jagdtigers were in the new operations area. Eight still needed minor repairs while 2 others were still in transport from St. Ingbert.

Once in the Haguenau forest the Jagdtigers were heavily camouflaged. The troops dug-in and prepared defensive firing positions, supplies were taken on and general preparations made.

32.7 Order from Insp. d. Pz. Tr. Berlin 16 February 1945

- - KR- - HZPH 018225/26 17.2.1650 AN HOKW: TM 2 =

NACHR OBDE / AHA / STAB = NACHR. INSP. D. PZ. TR. BERLIN =

Secret command matter.

Reference - Chief OKW / Heers Stab (ORG) Nr. 3432/45 G.K.V. 15.2.45

Business: Warning readiness s.Pz.Jg.Abt. (Jagdtiger) 512.

Operation of the Jagdtigers from s.Pz.Jg.Abt.653 acquired serious damage in the steering gear housing (constructive weakness).

It is necessary to change the unit on all Jagdtigers.

Design changes at first, to be made on all Jagdtigers that are still in production in the factory:

5 Jagdtigers of the s.Pz.Jg.Abt.512 in Döllersheim.
6 Jagdtigers in the factory in Linz.

The use of the s.Pz.Jg.Abt.512 will be delayed and you will be told the time and date of the design changes when it is known.

Gen. Insp.d.Pz.Tr./Lt.Kf.Offz.
Abt. Org. ROEM 2 Nr 746/45 G.K.V. 16.2.45
The Chief of Staff
Generalmajor Thomale.

As can be seen from this order, priority was given to the Jagdtigers of s.Pz.Jg.Abt.512

32.8. Changing the steering-units

This was one of the most difficult jobs for the Works Company. A mobile crane had to be positioned over the front of the vehicle, the gun assembly (8 metric tons [3]), was removed from its support mounts and swung out of the front using the lifting lug on the Saukopf mantle. The next stage was to remove the front deck plate to allow access to the top of the gearbox. The steering-unit was unbolted from the final drives and the prop-shaft withdrawn from the gearbox; special manipulator arms were fitted to the gearbox, which allowed it and the steering-unit to be swung out of the front deck plate. The transmission unit was placed on a clean surface and then the steering-unit removed from the gearbox. After removal the beveled output shaft and gearbox were checked over.

A new steering-unit was fitted and the whole process reversed. Lastly it was oiled up. After completion the gun had to be test fired to re-calibrate it for alignment with the optics. This was a long and difficult job (4 days), made worse by working outside in cold, wet conditions under large tarpaulins and camouflage nets.

Plate 253. Jagdtigers on rail transport moving to the assembly area (Karlheinz Münch).

Plate 254. Close up of one of the Jagdtigers (Karlheinz Münch).

Kopie   Archiv Karlheinz Münch

+--KR-- HZPH 018225/26 17.2.1650= AN HOKW : TM 2 =
NACHR.: OBDE/AHA/STAB =NACHR./: INSP.D.PZ.TR., BERLIN=
--GEHEIME KOMMANDOSACHE --.-
--BEZUG:-- CHEF OKW/HEERESSTAB (ORG) NR.3432/45 G.K.V.15.2.45.-
--BETR.:-- VERWENDUNGSBEREITSCHAFT S.H.PZ.JG.ABT.(JAGDTIGER) 512.- 1.) BEI EINSATZ DER JAGDTIGER DER S.H.PZ.JG.ABT.653 HAT SICH SERIENMAESSIGER SCHADEN AM LENKGETRIEBE ERGEBEN(KONSTRUKTIVE SCHWAECHE).- 2.) FORMAENDERUNG BEI ALLEN JAGDTIGERN ERFORDERLICH.- 3.) DURCHFUEHRUNG DER FORMAENDERUNG ERFOLGT ZUERST BEI JAGDTIGER DER NEUFERTIGUNG IM HEIMATKRIEGSGEBIET.- A) 5 JAGDTIGER DER S.H.PZ.JG.ABT.512 DOELLERSHEIM.- B) 6 JAGDTIGER IM H.ZA.LINZ.- 4.) HIERDURCH VERWENDUNGSBEREITSCHAFT 1./512 VERZOEGERT.- ZEITPUNKT DER VERWENDUNGSBEREITSCHAFT WIRD MITGETEILT, SOBALD ZEITBEDARF DER FORMAENDERUNG FESTSTEHT.= DER GEN.INSP.D.PZ.TR./LT.KF.OFFZ. ABT.ORG.ROEM 2 NR.746/45 G.K.V.16.2.45.- DER CHEF DES STABES GEZ.: THOMALE, GENERALMAJOR+

+1700 EIN KR TM 2 REHFELD HOKW +

Engine changes were much easier and quicker as far better access was afforded into the engine deck. None of the Jagdtigers belonging to 653 had the welded mounting brackets for fitting the 2 metric ton crane.

After repair the mechanics did the test drive to check out the vehicle. Once a clean bill of health was given, an entry was made in the vehicle's service book.

Each Jagdtiger was issued with a service book for entry of repairs and services to the vehicle. The Works Unit only used chassis number identification.

32.9 Proposed transfer of Porsche Jagdtigers

On 22 February, 653 receive the order (K.st.N.1148) to transfer their remaining six [4] *Porsche Jagdtigers to s.Pz.Jg.Kp.614.*

The s.Pz.Jg.Kp.614 were now under strength with only ten remaining Elefants. This order was not carried out due to the repair situation. A further record to s.Pz.Jg.Kp.614 on 3 March 45 still promised the six Jagdtigers (see Chapter 36).

On 28 February, a telegram from Gen Major Zimmerman (No 505/45) stated:

Battle command at St. Ingbert did not retain the 2 Jagdtigers; rail loading will take place at latest on 27.2.45 under the responsibility of the 653 battalion.

By 1 March, these had not arrived to the battalion and were the reason for the shortfall of 2 Jagdtigers in the strength report. (See Chapter 33.2)

[1] These are the last two Jagdtigers assigned to s.Pz.Jg.Abt. 653

[2] From the report on 5th February, there were no Jagdtigers in action, hence this must be the notification of the loss of No 134 at Rimling, to further substantiate this, and the log on 1st March also shows only this single Jagdtiger loss!

[3] 8 metric tons was over the limit of the 6 metric ton safe working load, of the Battalions two Sd. Kfz. 9/1 (Half-track cranes).

[4] This order indicates that 653 did not receive all seven of the Porsche Jagdtigers through the Ersatzheer in Döllersheim as stated in the record of 25 December 1944. From the report dated 16 January 1945, it is easy to conclude that either 305006, 305007 or 305008, was missing from the inventory of 653. To further substantiate this, the Nibelungen Werk release figures record a total of 36 Henschel Jagdtigers delivered to 653 between 6 October 1944 and 13 January 1945. The missing Porsche Jagdtiger was most likely returned to Nibelungen Werk from Döllersheim after a major mechanical failure, no records have been found to verify this!

— 40 —

Tag	Unterschrift und etwaige Bemerkungen des Prüfenden	Tag	Unterschrift und etwaige Bemerkungen des Prüfenden

Begleitheft

für

Kraft-wagen/rad W- [illegible]

Pz. Jg. Tiger B

305023

31 Kabzck Druck Wien 8., Wickenburggasse 13.

— 2 —

Beachte:

1. Halte das Begleitheft so sauber, wie es Dir möglich ist, schütze es vor Nässe.
2. Nicht nur der Abschnitt 2 ist wichtig für Dich, lies auch die anderen Abschnitte; sie machen Dich mit Deinem Kfz. vertraut.
3. Du darfst im Begleitheft nicht streichen, ändern, radieren; Unstimmigkeiten hast Du zu melden.
4. **Den Abschnitt 4, Ölwechsel,** mußt Du führen. Denke daran, daß diese Eintragungen dienstliche Meldungen sind. Sorgfältig durchgeführter Ölwechsel und sachverständiges Einfahren sind für Kfz. genau so wichtig, wie für Dich gute Verpflegung.

— 3 —

1. Beschreibung

Bezeichnung des Kraftfahrzeuges:	Abgekürzte Bezeichnung:	Anf. Nr.	Verladeklasse für Eisenbahntransport
Pz.Jg.Tiger			[illegible]

Herstellungsjahr 1944

Auf dem Rahmen und Haspel eingeschlagene Kontroll-Nr.

Eigengewicht, betriebsfertig, kg 75200

Höchstgewicht, beladen, kg

Laderaum: Länge m 7,80 (ohne Rohr); Breite m; Höhe m ~~10,37~~

Größte Außenmaße: Länge m 10,37; Breite m 3,67; Höhe m 2,95

Art und Größe der Bereifung: vorn; Mitte; hinten

Durchschnittsgeschwindigkeit km Std.: Straße; Gelände

Fassungsvermögen an: Kraftstoff, Liter 860; Öl im Motor, Liter 38

Kraftstoffverbrauch bei 100 km Fahrt in Litern: Gelände ca 1000 ltr.; Straße ca 800 ltr.

Fahrbereich mit einer Kraftstoffüllung: Straße km 100; Gelände km 70

Art des Motors: Vergaser, ~~Diesel~~; ~~Zweitakt~~ Viertakt

Art des Antriebes: Vorderrad; Hinterrad; Vierrad; Allrad; Gleiskette

Kraftübertragung: Kette; Kardan

Art der Kupplung

Zylinderinhalt cm^3 23000

Normaldrehzahl, Minuten 2800

1*

— 4 —

Magnet Zündstromerzeuger[1])	Fabrikat	Bosch
	Type	
Einstellung des Zündmoments	Hand	
	selbsttätig	
Vergaser	Fabrikat	Solex
	Type	52 JFF II
Lichtmaschine[1])	Fabrikat	Bosch
	Type	
Sammler	Fabrikat	
	Type	
	Ampèrestd.	
Anlasser	Fabrikat	
	Type	
Art der Kühlung	~~Luft~~	
	Wasser	
	mit Pumpe	
	~~ohne Pumpe~~	

Art der Ölung ....

Art der Kraftstoffzufuhr ....

Art u. Ausführung der Lenkung (Zweirad, Vierrad, Roßlenkung usw.) ....

Bremsen:

1. Feststellbremse wirkt auf ....
2. Betriebsbremse
   a) Art (Luftdruck, Öldruck, Unterdruck usw.) ....
   b) Fabrikat ....
   c) wirkt auf ....
   d) Höhe des Bremsdruckes bei Luftdruckbremse ....

Stoßdämpfer: Fabrikat .... Art ....

Art der Räder oder Felgen ....

Felgengröße ....

Art des Antriebes der Seilwinde oder Spills ....

Wärmewert der Zündkerzen[2]) ....

Luftdruck: vorn .... Mitte .... hinten ....

[1]) Zündlichtmaschinen nur unter Zündstromerzeuger eintragen, vorher diese Bezeichnung ändern.

[2]) Es sind die gleichen Wärmewerte mehrerer Fabrikate einzutragen.

— 5 —

Motor:

Maybach HL 230 TRM

n = 2800 U/min 720 PS

Verladebreite: 3,27 m.

Bodenfreiheit: 0,46 "

Spurweite: 2,79 "

Kettenbreite: 0,80 "

Feuerhöhe: 2,17 "

*Steigfähigkeit: 30°

Waat- " : 1,70 "

Kletter- " : 0,80 "

Graben-überschreitfähigkt: 1,80 "

Motor Nr. – 228CFG – 8324458

2

— 6 —

2. Nachweis der Ausrüstung

Anmerkung: Die Ausrüstung, die über das nach den zuständigen K-Anlagen festgesetzte Soll hinaus vorhanden ist, ist in den Spalten 1 und 3 handschriftlich nachzutragen.

Benennung	Soll*)	Istbestand	Bemerkungen
1	2	3	4
a) Zubehör			
Abblendkappe			
Abstützvorrichtung für Vorder- und Hinterfeder			
Anlaßhelf (o)		1	
Aschebehälter (o) (nur für Kfz. mit handelsüblichem Aufbau)[1])			
Begleitsitz (o)			
Beleuchtungseinrichtung: Suchlampe (o) mit Leitungskabel			
Handlampe mit Schutzkorb, 3 m Kabel und Stecker (o)			
Innenbeleuchtung (o) (nur für geschlossene Kfz.)			
Seitenlampe (o)			
Scheinwerfer mit Halter (o)			
Schlußlampe mit Halter (o)			
Spannungsmesser (Voltmeter) (o)			
Steckdose (o)			
Beutel aus Drillich, 220 × 120 mm, mit Schnur (für Putzwolle oder Putzlappen)			
Bindestrang, 2,5 m lang			
Bohle aus Hartholz, mit Blech beschlagenen, abgeschrägten Enden: 100 × 40 × 5 cm			
178 × 45 × 5 cm			
Brechstange, 1200 mm lang			
Brechstange, 1800 mm lang		1	
Brückenschiene mit 2 Spur- und 1 Verbindungsstange			
Büchsen bzw. Behälter für Düsen			
für Kraftradschlauch (o)			

[1]) Bei Kraftomnibussen je Fenster 1 Aschebehälter.

*) In diese Spalte ist von der ausstellenden Einheit usw. die nach den entsprechenden K-Anlagen und Sonderbestimmungen zustehende Ausrüstung für das betr. Kfz. in Rot einzutragen.

— 7 —

Benennung	Soll*)	Istbestand	Bemerkungen
1	2	3	4
Noch: a) Zubehör			
Blechbüchse, rund, mit Klemmdeckel, 80 mm Ø, für ~~Federringe, Unterlegscheiben, Muttern, Schrauben~~, Splinte		1	
Noch: Büchsen bzw. Behälter aus Blech 45 mm Ø für Schmirgel			
80 mm Ø für Staufferfett			
56 mm Ø für Staufferfett			
45 mm Ø für Talkum (auch Streuer zulässig) (nur bei Luftbereifung)			
Decken für Insassen (Pelzdecke 1,5 × 2,0 m (o) (nur für offene Pkw.)			
für Insassen (Wolldecke mit Ledereinfassung) (o) (nur für geschlossene und für l. u. m. offene Pkw.)			
für Kühler (Kühlerschutzhaube) (o) (aus Kunstleder)			
Druckluftmesser (o)			
Fahrtrichtungsanzeiger (o) (beleucht- und herausklappbar)			
Frostschutzscheibe (o) (nur für geschlossene Pkw. und Krtw.) (o)			
Fußrasten für Krad (o)			
Gepäck- oder Gerätekoffer (o)[1])			
Gepäckhalter mit 2 Riemen (o)			
Geschwindigkeitsmesser mit Kilometer- und Tageszähler (o)			
Gleitschutz: Geländekette			
Gleitschutzkette (o)			
Kettenspanner (o) für Gleitschutzkette			
Schneekufe (o)			

[1]) Soweit bei Neulieferung vorhanden.

*) In diese Spalte ist von der ausstellenden Einheit usw. die nach den entsprechenden K-Anlagen und Sonderbestimmungen zustehenden Ausrüstung für das betr. Kfz. in Rot einzutragen.

2*

— 8 —

Benennung	Soll *)	Istbestand	Bemerkungen
1	2	3	4
Noch: a) Zubehör			
Heizeinrichtung für geschlossene Kraftfahrzeuge (o) 2)			
Hupe mit Gummiball oder elektrisch, mit Knopf oder Signalring (o) 1)			
Kannen bzw. Behälter			
Kanister, dreikantig, explosionssicher für 5 l (für Kraftstoff)			
für 10–15 l (für Kraftstoff) 2)			
Kanister, vierkantig, explosionssicher für 1 l (für Bremsflüssigkeit) 2)			
für 2–5 l (für Heißdampfzylinderöl)			
für 2–5 l (für Motorenöl)			
für 0,5 l (für Petroleum)			
Handölkanne, eiförmig, für 0,25 l		1	
Spritzkanne, vierkantig, mit geradem Rohr und Nadel für 0,08 l für Kraftstoff			
für Petroleum			
für Motorenöl			
Karabinerhalter			
Kasten für Gepäck (o)			
für Gleitschutz und Kannen (o) (mit Zwischenwand und Abflußlöchern im Boden)			
für Sand (o) (nur für vollgummi- und elastikbereifte Kraftfahrzeuge)			
für Zubehör und Vorratssachen (o)			
Ketten für Plane (o)			
Sicherheitskette oder Bügel f. Wagenwinde (o)			
Spannkette aus 2 Hälften (o) für Wagenkasten			

1) Funkkraftfahrzeuge erhalten beide Signalvorrichtungen. — 2) Soweit erforderlich.
*) In diese Spalte ist von der ausstellenden Einheit usw. die nach den entsprechenden K-Anlagen und Sonderbestimmungen zustehende Ausrüstung für das betr. Kfz. in Rot einzutragen.

— 9 —

Benennung	Soll *)	Istbestand	Bemerkungen
1	2	3	4
Noch: a) Zubehör			
Motorpfeife mit Bowdenzug und Betätigungshebel (o) (für Pkw. nur, wenn durch die Motorbauart möglich)			
Packtaschen mit Bezug und Tragegestell (o) Paar			
Packtaschen mit Bezug (o) Paar			
Plane ~~mit Spriegel (o)~~ für Grating		2	
Plane für ~~Gepäckgalerie (o)~~ Kampfraum		1	
Polsterkissen			
Polsterkissen für Rückenlehnen (ausschl. Führer- und Begleitersitze)			
Polsterkissen für Sitzbänke (ausschl. Führer- und Begleitersitze)			
Regenschutzkappe (o) für Zündkerzen			
Reifenhalter, verschließbar, für Vorratsbereifung (o)			
Rückblickspiegel (o)			
C ~~S~~-Haken 5000 kg Tragfähigkeit ~~(für Seilwinde)~~		2	
Scheibenwischer (o)			
Schießklotz mit Vorsteckbohle			
Schild für polizeiliche Kennzeichen (o)			
Schlüsselring mit Schild		1	
Schutzhülle aus Leder (für Seilwickelvorrichtung)			
Segeltucheimer mit Ausgußtülle für 11 l		1	
Seilrolle, lose (o)			
Signalvorrichtung, mechanisch, bestehend aus:			
Alarmglocke (am Führersitz des Kraftfahrzeuges befestigt)			
Zugleine 12 m lang, mit Karabinerhaken			
Sitzbrett 35 cm breit			
Stab zum Zeichengeben (o)			
Stahldrahtseile			
passend zum Spill, 20 m lang (o)			
auf Trommel aufgewickelt (o) ..... m			

*) In diese Spalte ist von der ausstellenden Einheit usw. die nach den entsprechenden K-Anlagen und Sonderbestimmungen zustehende Ausrüstung für das betr. Kfz. in Rot einzutragen.

— 10 —

Benennung	Soll *)	Istbestand	Bemerkungen
1	2	3	4
Noch: a) Zubehör			
Noch: Stahldrahtseile			
auf Seiltrommel, 16 mm Ø (o) .... m			
mit Ring ~~und Haken~~ 8,2 m lang, 32 mm Ø		2	
mit Ring ~~u. Haken~~ 5 m lang, etwa 14 mm Ø		1	
Stoßstange (o)			
Taschen			
mit Lederrand (o) für Ausweise usw.		1	
aus Segeltuch oder Leder (o), für Werkzeug			
Trichter			
oval, 95×65 mm Ø (für Motorenöl)			
oval, 125×80 mm (für Kraftstoff)			
Uhr für Kraftfahrzeuge (o) (wenn bei Beschaffung des Kfz. mitgeliefert)			
Unterlegklotz für Wagenwinde (o)		1	
Verbindungskabel 2,5 m lang, mit 2 Steckern für den Anschluß zum Zugwagen (o)			
Vorhängesicherheitsschloß mit 2 Schlüsseln für Kästen am Kfz.			
für Sicherheitskette oder Bügel zur Wagenwinde			
für Plane			
Windschutzscheibe (o)			
Werkzeug			
Aufzieheisen (Montierhebel) (o) f. Luftbereifung			
Abzugglocke für Kegelrad			
Abzugschraube ½" Gasgewinde, 268 mm lang			
Abzugbügel für Antriebsrad zur Wasserpumpe			
Abzugschraube ½" zum Abzugbügel			
Abzugbügel für Lenkrad			
Abzugschraube ⅝" zum Abzugbügel für Lenkrad			

*) In diese Spalte ist von der ausstellenden Einheit usw. die nach den entsprechenden K-Anlagen und Sonderbestimmungen zustehende Ausrüstung für das betr. Kfz. in Rot einzutragen.

— 11 —

Benennung	Soll *)	Istbestand	Bemerkungen
1	2	3	4
Noch: a) Zubehör (Werkzeug)			
Abzugglocke mit Schraube für konisches Zahnrad auf der Schneckenwelle			
Abzugschraube 2½" R für die Abzugglocke			
Abzugteller zur Sicherungsmutter für die Achsgabel am Tragrohr			
Abzugmuffe für Vorderrad mit Sechskant			
Abzugmuffe für Hinterrad mit Sechskant			
Abdruckschraube für S-Felge			
Abzugschraube für Hinterrad			
Bürsten			
Polsterbürste (für geschlossene Pkw., Innenpolsterung mit Stoffbezug)			
Wagenwaschbürste mit Griff			
Zündkerzenbürste mit Griff		1	
Druckstück zum Abzugbügel			
Durchtreiber (Durchschlag), rund			
3 mm Ø			
6 mm Ø		2	
10 mm Ø			
Feilen mit Heft 1)			
Flachstumpffeile, 300 mm lang, Bastard			
Halbrundfeile, 150 mm lang, Bastard m. Heft		1	
Halbrundfeile, 200 mm lang, Bastard			
Rundfeile, 150 mm lang, Bastard			
Vierkantfeile, 200 mm lang, Bastard			
Handfeilkloben mit breitem Maul, 100 mm lang			
Hebel zum Einhängen der Federn für die Hinterradbremsen			
Hilfsbolzen zur Einführung der Hinterradbremsen			
Meißel			
Flachmeißel, 100 mm lang			

1) In je eine Segeltuchhülle verpackt.
*) In diese Spalte ist von der ausstellenden Einheit usw. die nach den entsprechenden K-Anlagen und Sonderbestimmungen zustehende Ausrüstung für das betr. Kfz. in Rot einzutragen.

— 12 —

Benennung	Soll*)	Istbestand	Bemerkungen
1	2	3	4
Noch: a) Zubehör (Werkzeug)			
Noch: Meißel Flachmeißel, 200 mm lang		1	
Kreuzmeißel, 150 mm lang			
Nietenlöser (o) für Antriebskette (Krad)			
Pumpen Druckluftpumpe[1]), eingebaut und mechanisch angetrieben, mit Pumpenschlauch, 4,5 m lang, Druckanzeiger und Verschraubung zum Anschluß an die Pumpe (o)			
Handluftpumpe mit Schlauch und Druckanzeiger (o)			
Rad- bzw. Ritzelabzieher (o)			
Riemenlocher (o) (soweit erforderlich)			
Schlosserhammer 500 g schwer		1	
1500 g schwer		1	
800 g schwer			
Schlüssel für Achsmutter			
für Kraftstoffbehälter (soweit erforderlich)			
für Magnet (soweit erforderlich)			
für Radkapsel (soweit erforderlich)			
für Rudgeräder (soweit erforderlich)			
für Radbolzen (soweit erforderlich)			
für Speichen			
für Vergaser			
Doppelschraubenschlüssel (Chrom-Vanadium)		5	
Kurzschlußschlüssel (o)			
Ringmutterschlüssel 20×33×44 mm			
Schraubenschlüssel, verstellbar 125 mm lang		1	

[1]) Nur wenn bei Beschaffung des Kfz. mitgeliefert.
*) In diese Spalte ist von der ausstellenden Einheit usw. die nach den entsprechenden K-Anlagen und Sonderbestimmungen zustehende Ausrüstung für das betr. Kfz. in Rot einzutragen.

— 13 —

Benennung	Soll*)	Istbestand	Bemerkungen
1	2	3	4
Noch: a) Zubehör (Werkzeug)			
Noch: Schlüssel 206 mm lang			
240 mm lang			
Sechskantsteckschlüssel aus Rohr, doppelseitig		5	
Drehstift für Sechskantschlüssel (Stufen)		1	
Sonderwerkzeuge			
Steckschlüssel für Kupplungsschrauben zum Antrieb der Wasserpumpe			
Steckschlüssel mit Sechskant für Nippel am Vorderradgehäuse			
Steckschlüssel mit Sechskant für die Nippel zum Ölstand am Wechselrädergehäuse			
Steckschlüssel 22 mm, für Zylinderbefestigung			
Steckschlüssel für Ventilverschraubung oder Deckel			
Steckschlüssel für Gelenkkupplung			
Zapfenschlüssel für Lenkung			
Zapfenschlüssel f. kon. Rad am Vorderradantrieb			
Zapfenschlüssel für Vorderrad und Vorderachsschenkel			
Zapfenschlüssel für Überwurfmutter zur Wasserpumpe und Stopfbüchse			
Schraubenzieher mit durchgehender Klinge 150 mm lang, 5 mm Schnittbreite		1	
300 mm lang, 12 mm Schnittbreite		1	
~~Spritzen~~ Lub- ~~Hochdruck~~fettpresse (o) (soweit erforderlich)		2	
Handfett- ~~Hochdruck~~ölpresse (o) mit Schlauch		1	
Wagenheber, mechanisch, mit Schlüssel u. Griff (o) (Tragfähigkeit soll dem Eigengewicht des Kfz. entsprechen)			
Wagenwinde mit Kurbel (Tragfähigkeit soll dem Gewicht des beladenen Kfz. entsprechen)		1	
Waschpinsel			

*) In diese Spalte ist von der ausstellenden Einheit usw. die nach den entsprechenden K-Anlagen und Sonderbestimmungen zustehende Ausrüstung für das betr. Kfz. in Rot einzutragen.

3

— 14 —

Benennung	Soll*)	Istbestand	Bemerkungen
1	2	3	4
Noch: a) Zubehör (Werkzeug)			
Brechbaum mit Ring, Griff und abgerundetem Eisenschuh (o)			
Zapfen zur Abzugsmuffe für Vorderrad			
für Hinterrad			
Zangen Kombinationszange, 160 mm lang		1	
Gaszange, 225 mm lang			
Vorschneider, 160 mm lang			
b) Vorratssachen			
Anschlußnippel für Hochdruckfettpresse (o) (soweit erforderlich)			
Antriebskette oder -riemen (o)			
Ausgleichscheibe für Hinterrad			
Behälter aus Glas (zum Ersatz) für die Brennstoffförderungsvorrichtung (soweit erforderlich)			
Bereifung (o) Decke (nur bei Luftbereifung, auf Vorratsfelge oder -rad aufgezogen)			
Schlauch (nur bei Luftbereifung, auf Vorratsfelge oder -rad aufgezogen)			
Vollgummireifen			
Bowdenzug (o)			
Druckscheibe für Achsschenkel (unten)			
Düsen für Vergaser (o) Satz			
Endglieder für Gleitschutzkette (o)			
Felge oder Rad f. luftbereifte Kraftfahrzeuge (o)			
Kasten mit Vorratssachen für Zünd-, Licht- und Anlasseranlagen (o) (Inhalt nach Fabrik und Type)			
Kettenglieder für Kraftrad gerade (o) (bei Kettenbetrieb)			
gekröpft (o) (bei Kettenbetrieb)			

*) In diese Spalte ist von der ausstellenden Einheit usw. die nach den entsprechenden K-Anlagen und Sonderbestimmungen zustehende Ausrüstung für das betr. Kfz. in Rot einzutragen.

— 15 —

Benennung	Soll*)	Istbestand	Bemerkungen
1	2	3	4
Noch: b) Vorratssachen			
Noch: Kettenglieder für Laufkette			
für Gleitschutzkette mit 2 Verbindungsgliedern			
für Sechsradkette			
Kettengliedbolzen			
Sechsradkettenbolzen			
Ventile mit Zubehör Auslaßventil			
Einlaßventil			
Werkstoffe			
Dichtungsring (Fiber)			
Dichtungsschnur, 110 mm lang, 8 mm ∅			
Eisendraht (Bindedraht) kg		0,1	
Gummi-Instandsetzungsmittel für luftbereifte Kraftfahrzeuge (in Kasten oder Büchse)			
Isolierband, 25 m lang, 16 mm breit ~~(in Blechbüchse)~~ Büchse		1	
Packungen (Dichtungen) nach Fabrikat (in besonderem Behälter) Satz		9	
Riemen für Windflügel (bei Riemenantrieb)			
Riemenverbinder (bei Riemenantrieb)			
für Antriebsriemen (bei Riemenantrieb)			
für Windflügelriemen (bei Riemenantrieb)			
Schlauch für Kühler oder Wasserpumpe			
Schlauchbinder			
Schrauben, Muttern, Splinte, Unterlegscheiben, Federringe			
Schrauben für Vorder- und Hinterradfelge			
Ventileinsatz für Luftbereifung Schachtel			
Zündkabel m			
Zündkerzen ~~in Schutzhülle~~ mit Dichtring		3	

*) In diese Spalte ist von der ausstellenden Einheit usw. die nach den entsprechenden K-Anlagen und Sonderbestimmungen zustehende Ausrüstung für das betr. Kfz. in Rot einzutragen.

3*

Benennung	Soll *)	Istbestand	Bemerkungen
1	2	3	4
Dem Gerät beigegebene Vorschriften und Anweisungen			
Beschreibung und Bedienungsvorschrift für das Kraftfahrzeug			
Beschreibung und Bedienungsvorschrift für Zündanlage			
für Lichtanlage			
Begleitheft ~~(mit Tintenstift)~~		1	
Ersatzteilliste			
Kraftfahrzeugschein			
Kiste f. Ers.Teile u.Werkzeug			
Putzlappen kg.		0,5	
Steckschlüssel für Lauf-u.Leitrad SW 105		1	
Schraubenschlüssel 55/85		1	
Doppelringschlüssel 30/32		1	
Öleinfülltrichter		1	
Sicherungen 15 Amp.		10	
" 40 "			
Reserveteilkasten mit:		1	
Stellringe f.Kettenbolzen			
Sechskantschrauben mit		24	
Muttern		24	
Stoßeinsatz			W u.G

*) In diese Spalte ist von der ausstellenden Einheit usw. die nach den entsprechenden K-Anlagen und Sonderbestimmungen zustehende Ausrüstung für das betr. Kfz. in Rot einzutragen.

Benennung	Soll *)	Istbestand	Bemerkungen
1	2	3	4
Rändelschrauben mit		18	
Muttern		18	
Sicherungsbleche		12	
Kasten f.Mannsch.-Gepäck		1	
Kettenschließer		1	
Kettenspannschlüssel		1	
Heizlampe 2 Ltr.		1	
Vortreiber f.Kettenbolzen		1	
Zündschlüssel		1	
Doppelkettenglieder		6	
Kettenbolzen		10	
Vorschlaghammer 6 kg.		1	
Hand- ~~Magnet~~lampe		1	
Lukenschlüssel		2	
Arkarol Pack.			
Prismengläser		20	
Motor-Werkzeugkasten:		1	
Ventileinstellehre		1	
Düsen/Zündkerzenschlüssel		1	
Drehstift		1	

*) In diese Spalte ist von der ausstellenden Einheit usw. die nach den entsprechenden K-Anlagen und Sonderbestimmungen zustehende Ausrüstung für das betr. Kfz. in Rot einzutragen.

Benennung	Soll *)	Istbestand	Bemerkungen
1	2	3	4

*) In diese Spalte ist von der ausstellenden Einheit usw. die nach den entsprechenden K-Anlagen und Sonderbestimmungen zustehende Ausrüstung für das betr. Kfz. in Rot einzutragen.

Benennung	Soll *)	Istbestand	Bemerkungen
1	2	3	4
21. Kraftfahrzeuge (Zubehör)			
Fahnenstange[1]) mit je einer roten, weißen und blauen Flagge als Parteiabzeichen			
8 Drahtgitter aus vernickeltem Messingdraht (als Scheibenschutz) . . . (nur für Krankenkraftwagen)			
1 Kasten, tragbar (zugleich Sitzbank mit 2 Kissen) . . . . (nur für Krankenkraftwagen)			
1 Kokosmatte (o) für Innenraum . (nur für Krankenkraftwagen)			
1 Leiter aus Holz, zusammenlegbar, etwa 2,50 m lang . . . . . (nur für Krankenkraftwagen)			
1 Rückenlehne . . . . (nur für Krankenkraftwagen)			
1 Sitzbank für 4 Personen mit umklappbaren Fußstützen . . . . . . (nur für Krankenkraftwagen)			
4 Traggestelle für Krankentragen . . (nur für Krankenkraftwagen)			
1 Trinkwasserbehälter, 15 l Inhalt, mit Absperr-Rohr und 2 Anschnallriemen . . . . (nur für Krankenkraftwagen)			
29. Schanzzeug			
Axt, lange		1	
Drahtschere, kleine		1	
Klauenbeil			
Kreuzhacke lang			
~~schwer~~ Schaufel		1	
Spaten lang			
halblang			

[1]) Ausrüstung mit Fahnenstange gilt für a l l e Kraftfahrzeuge. Für Krafträder treten an Stelle der Fahnenstange Vorrichtungen gem. Nw. M. H. L. II D Nr. 109. 8. 27. In. 6 v. 18. 8. 27.
*) In diese Spalte ist von der ausstellenden Einheit usw. die nach den entsprechenden K-Anlagen und Sonderbestimmungen zustehende Ausrüstung für das betr. Kfz. in Rot einzutragen.

— 20 —

Benennung	Soll*)	Istbestand	Bemerkungen
1	2	3	4
33. Feuerlöschgerät			
Schaumlöscher, 10 l, frostsicher[1])			
Trockenfeuerlöscher, 4 kg, mit Aufhängevorrichtung[2])			
Naßfeuerlöscher[2]) Tetra		2	
36 b. Krankenpflege und Transportgerät			
4 zusammenlegbare Krankentragen mit 2 Gurten . . . . (nur für Krankenwagen)			
8 wollene Decken (o) (zur Krankenpflege) . . . . (nur für Krankenwagen)			
36 f. Sanitätsausrüstungseinheiten gemischten Inhalts			
Verbandkasten mit Inhalt[3])		1	
47. Betriebsstoff und Reinigungsgerät			
Lederlappen			
Putzwolle oder Putzlappen . . . . kg			
Schmirgel, fein . . . . kg			
Schmirgelleinwand, fein . . . . Bogen			
Schwamm			

[1]) Bei Beförderung von feuergefährlichen Stoffen (z. B. Kraftstoff, Öl, Putzwolle, Karbid, Munition, Sprengstoffe). Sonstige Zuteilung ist erforderlich an Stellen, wo eine besonders auf Entzündung von Benzin und ähnlichen Stoffen fußende Brandgefahr vorliegt und die Mitführung nach Raum und Gewicht möglich ist. — [2]) Lastkraftwagen und Anhänger, die feuergefährliche Stoffe (s. Fußnote 1) befördern, erhalten einen dieser Feuerlöscher neben dem zuständigen Schaumlöscher. — [3]) Verbandpäckchen für Kraftradfahrer siehe H. V. Bl. 1931 Ziffer 49.

*) In diese Spalte ist von der ausstellenden Einheit usw. die nach den entsprechenden K-Anlagen und Sonderbestimmungen zustehende Ausrüstung für das betr. Kfz. in Rot einzutragen.

— 21 —

Benennung	Soll*)	Istbestand	Bemerkungen
1	2	3	4

*) In diese Spalte ist von der ausstellenden Einheit usw. die nach den entsprechenden K-Anlagen und Sonderbestimmungen zustehende Ausrüstung für das betr. Kfz. in Rot einzutragen.

— 22 — — 23 —

3. Übernahme und Prüfnachweis über die Vollzähligkeit der Ausrüstung

Tag	Unterschrift, Dienstgrad und Truppenteil des Übergebenden	Unterschrift, Dienstgrad und Truppenteil des Übernehmenden	Vermerke über Vollzähligkeit, Fehlen und Ersatz des fehlenden Geräts, sonstige erforderliche Bemerkungen	Unterschrift, Dienstgrad und Name des Prüfenden oder die Übernahme Beaufsichtigenden
ab 21.11.		Dienststelle Feldpost Nr. 25056/C	Werkzeug u. Ausrüstung vollzählig	Henning Uffz.
4.12.44.	[illegible] Ogfr.	[illegible]	Ausrüstung vollzählig	[illegible] Ogfr.
15.1.45	[illegible]	[illegible]	[illegible]	
			[illegible]	Henning Uffz.

— 24 —

Tag	Unterschrift, Dienstgrad und Truppenteil des	
	Übergebenden	Übernehmenden

— 25 —

Vermerke über Vollzähligkeit, Fehlen und Ersatz des fehlenden Geräts, sonstige erforderliche Bemerkungen	Unterschrift, Dienstgrad und Name des Prüfenden oder die Übernahme Beaufsichtigenden
Schaltgetriebe erneuert	am 13.II.45
Motor ausgewechselt	" 18.II.45
Laufwerk generalüberholt	" 3.4.45
Leitrad erneuert	" 6.4.45
Schaltgetriebe ausgewechselt	" 8.4.45
Vergaser ausgetauscht	" 10.4.45
Bremsen überholt	" 12.4.45
Kanone überprüft	" 3.[illegible]45
Lenkgerät erneuert	" 24.4.45
Lüftergitter erneuert	" 25.4.45
Drehstab ausgewechselt	" 28.3.45
Lenkbremse rechts erneuert	" 30.3.45
Kleine [illegible]	

33

Schwere Panzerjäger Abteilung 653 March 1945

33.1 March 1945 combat synopsis

1 March There are 39 Jagdtigers in the unit's assembly area in the Haguenau Forest, of which 8 are in short-term repair. The Battalion is the armored reserve behind the Moder line. The Works Company is working on 2 Jagdtigers in long-term repair at Bellheim. The 653 are located in the Schirlenhoff-Surbourg area.

6 March Private Herman Looft 1/653 receives the Tank Fighting Medal in Silver for being keen and brave on 25 occasions.

7 March US troops capture the Rail Bridge at Remagen and cross the Rhine.

11 March Over the 11 days, 7 Jagdtigers have been repaired and sent by rail to Surbourg from the repair area. The 653 reach their maximum strength of 41 Jagdtigers with 38 operational.

14 March 3/653 in night battle at Griesbach, American column is shot up.

15 March At 0700 hours French Colonial forces attack over the Moder River

16 March Moder line penetrated. 3/653 counter attack they are caught by air attack at Morsbronn: 5 Jagdtigers damaged by rockets, 2 more by artillery fire, only 2 could be retrieved 5 are blown up. The broken down vehicles withdrawn by rail that night, 7 Jagdtigers remain as rearguard.

17 March French break through at Oberhoffen after 36-hour attack, Haguenau left flank is breached. The 7 Jagdtigers counterattack the American spearhead at Hegeney. The 1/653 has to blow up Jagdtiger (102). Near Rittershoffen.

18 March Strength report shows 653 with 34 Jagdtigers; 18 operational with 16 needing repair. The 653 withdraw through Wissembourg behind the Siegfried Line. 2/653 lose Jagdtiger just before Wissembourg, it had to be blown up.

19 March 653 withdraw behind the Siegfried Line, they suffer numerous air attacks in the Wissembourg/Lauterbourg area. The French Colonial attack has great difficulty in the forest of the Bienwald.

20 March The 653 units group in the assembly area near Bellheim.

21 March Hitler orders that in all circumstances, 653 be conveyed safely across the Rhine. Unit at Schifferstadt, three Kampfgruppen deploy on the line Neustadt-Böhl-Schifferstadt.

22 March Americans breakthrough Siegfried Line at Wissembourg a Jagdtiger from 3/653 is blown up near Schweigen. The 3/653 smash a U.S. armored column near Neustadt am Der Weinstrasse. 1/653 and 2/653 also have successes on their right flank That night Patton succeeds in crossing the Rhine at Oppenheim. The 2/653 lose 2 Jagdtigers.

23 March Early morning, 2 Jagdtigers abandoned in Neustadt. Strength report shows 653 with 31 Jagdtigers, 3 fully operational. 28 vehicles had been taken across the Rhine. Berghausen/Germersheim is the unit's location. Unit relocates to woods near Neudorf. Grillinberger requests a 2-week stand-down of 653 for technical reasons. Jagdtiger 234 is blown up in Zeiskam.

24 March Bridge over the Rhine blown-up at Germersheim after 653 had crossed the river.

25 March The west bank of the Rhine is completely cleared of all German fighting forces. Major Grillinberger is demoted to Lieutenant for tactical blunders; he was also unpopular with the troops. Major Rolf Fromme the former Commander of s.Pz.Abt. 503 is appointed the new Commander. The s.Pz.Jg.Abt. 653 have a 2-day refit in the Neudorf area.

76

Meldung vom 4. März 1945

Verband: schw.Pz.Jäg.Abt.653
Unterstellungsverhältnis: XC.A.K.

1. Personelle Lage am Stichtag der Meldung: 1.3.45

a) **Personal:**

	Soll	Fehl	
Offiziere	29 u.1 Bea.	3	+)
Uffz.	272 u.4 Bea.		
Mannsch.	681	10	
Hiwi	13		
Insgesamt	1000	13	+)

b) **Verluste und sonstige Abgänge:**
in der Berichtszeit vom 1.2. bis 28.2.45

	tot	verw.	verm.	krank	sonst.
Offiziere	1	-	-	-	(vers.) -
Uffz. und Mannsch.	-	2	-	14	26
Insgesamt	1	2	-	14	26

c) **in der Berichtszeit eingetroffener Ersatz:**

	Ersatz	Genesene
Offiziere	1	-
Uffz. und Mannsch.	24	3

d) **über 1 Jahr nicht beurlaubt:**

insgesamt: 70 Köpfe 6,7 % d. Iststärke

davon:	12–18 Monate	19–24 Monate	über 24 Monate
	70	-	-
Platzkarten im Berichtsmonat zugewiesen:	-		

2. Materielle Lage:

		Gepanzerte Fahrzeuge							Kraftfahrzeuge				
		Stu.-Gesch.	III	IV	V	VI	Schtz. Pz. Pz. Sp. Art. Pz. B. (o. Pz. Fu. Wg.)	Pak. SF.	Kräder Ketten	Kräder m. angetr. Bwg.	Kräder sonst.	Pkw gel.	Pkw 0
Soll (Zahlen)		-	-	8	5	45	10	-	14	-	8	39	2
einsatzbereit	zahlenm.	-	-	6	3	31	14	-	-	2	13	32	2
	in % des Solls	-	-	75	60	70	140	-	-	-	163	82	100
in kurzfristig. Instandsetzg. (bis 3 Wochen)	zahlenm.	-	-	2	1	8	-	-	-	1	-	5	-
	in % des Solls	-	-	25	20	18	-	-	-	-	-	12	-

		noch Kraftfahrzeuge							Waffen			
		Lkw Maultier.	Lkw m	Lkw s	Lkw Tonnage	Ketten-Fahrzeuge Zgkw. *)	Ketten-Fahrzeuge Zgkw. **)	Ketten-Fahrzeuge RSO	s. Pak 80	2 cm Vierl	MG.	3,7 cm Fla
Soll (Zahlen		6	32	87	461	8	15	-	45	7	126(58)	4
einsatzbereit	zahlenm.	13	30	53	322	6	14	-	31	7	119(54)	3
	in % des Solls	217	94	61	70	75	93	-	70	100	94(92)	75
in kurzfristig. Instandsetzg. (bis 3 Wochen)	zahlenm.	3	2	9	52	-	3	-	8	-	-	1
	in % des Solls	50	6	10	11	-	20	-	18	-	-	25

*) Zgkw. mit 1–5 t, **) Zgkw. mit 8–18 t
( ) davon MG. 42

3. Pferdefehlstellen: - - -

+) 2 Fehlstellen für Offz. sind von 2 zur Abt. kommandierten Offz. ausgefüllt, 1 Hilfs-Arzt und 10 Mannschaften wurden angefordert.

Anl. zu Nr. 00 269 / 45 geh.

Wehrkreisdruckerei VIII, Breslau. 2220 I.

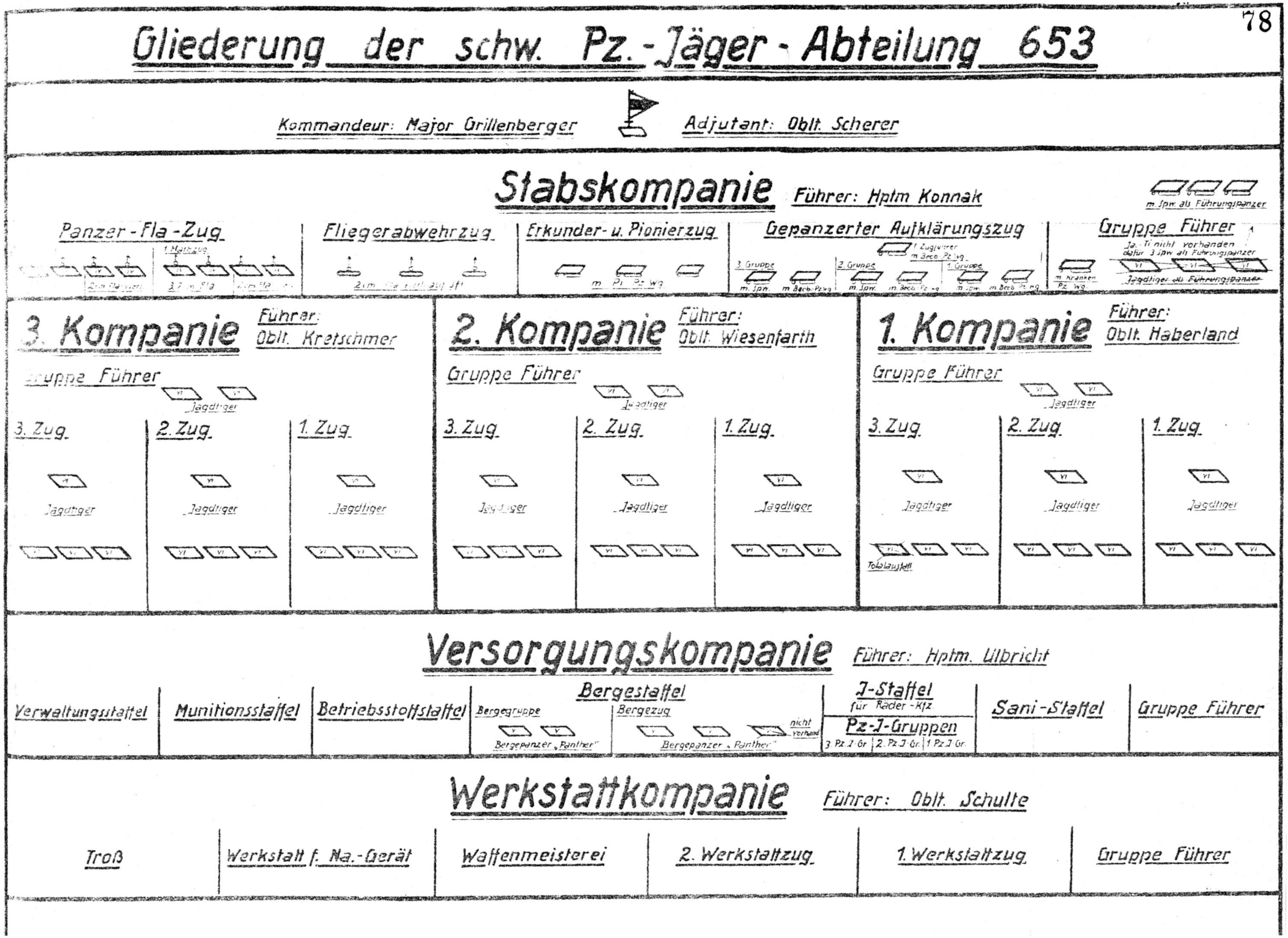

78

Gliederung der schw. Pz.-Jäger-Abteilung 653

Kommandeur: Major Grillenberger — Adjutant: Oblt. Scherer

Stabskompanie

Führer: Hptm Konnak

m. Spw. als Führungspanzer

Panzer-Fla-Zug	Fliegerabwehrzug	Erkunder- u. Pionierzug	Gepanzerter Aufklärungszug	Gruppe Führer
1. Halbzug; 3,7 cm Fla	2 cm Fla	m. Pi. Pz. Wg.	1 Zugführer m. Beob. Pz. Wg.; 3. Gruppe; 2. Gruppe; 1. Gruppe; m. Spw.; m. Beob. Pz. Wg.	m. kranken Pz. Wg.; Ja. Ti nicht vorhanden dafür 3 Spw als Führungspanzer; Jagdtiger als Führungspanzer

3. Kompanie

Führer: Oblt. Kretschmer

Gruppe Führer — Jagdtiger

3. Zug	2. Zug	1. Zug
Jagdtiger	Jagdtiger	Jagdtiger

2. Kompanie

Führer: Oblt. Wiesenfarth

Gruppe Führer — Jagdtiger

3. Zug	2. Zug	1. Zug
Jagdtiger	Jagdtiger	Jagdtiger

1. Kompanie

Führer: Oblt. Haberland

Gruppe Führer — Jagdtiger

3. Zug	2. Zug	1. Zug
Jagdtiger; Totalausfall	Jagdtiger	Jagdtiger

Versorgungskompanie

Führer: Hptm. Ulbricht

Verwaltungsstaffel	Munitionsstaffel	Betriebsstoffstaffel	Bergestaffel	J-Staffel für Räder-Kfz.; Pz-J-Gruppen	Sani-Staffel	Gruppe Führer
			Bergegruppe: Bergepanzer „Panther"; Bergezug: Bergepanzer „Panther", nicht vorhand.	3. Pz. J. Gr.; 2. Pz. J. Gr.; 1. Pz. J. Gr.		

Werkstattkompanie

Führer: Oblt. Schulte

Troß	Werkstatt f. Na.-Gerät	Waffenmeisterei	2. Werkstattzug	1. Werkstattzug	Gruppe Führer

26 March Out of the 28 Jagdtigers, 9 are operational, 19 need repair. Of these there are 10 long-term (5 days) repairs.

27 March The French 1st Army moves north across the Rhine near Oppenheim. 653 are ordered north towards Mannheim to counter the Allied bridgehead. 653 make a reconnaissance at Grenzhof.

28 March 653 reconnaissance at Grenzhof. Lt. Hans Knippenberg receives the Iron-Cross from Major Rolf Fromme.

29 March A Kampfgruppe with 6 Jagdtigers move forward towards Schwetzingen, 3 breakdown on route. The 10 Jagdtigers needing long term repair are sent by rail from Neudorf to the safer area of Stuttgart, with the Workshop Company.

30 March A strength report still shows 653 with 28 Jagdtigers; 6 operational, 22 needing repair, 3 Jagdtigers in Schwetzingen are lost 1 in combat and 2 blown up by the crews.

31 March The 3 operational Jagdtigers move towards Eppingen.

33.2 Report dated 4 /3 /45 s.Pz.Jg.Abt.653 (XC.A.K.)

Personnel log: a. Total of 1000 people; 29 officers, 271 NCO's, 681 men, 13 foreign vol'
1/2/45 to 28/2/45 b. Casualties and departure (1 killed, 2 wounded, 14 sick, 26 others)

c. In the promised list of replacements are 1 officer and 42 men.

Material log: Jagdtiger (quota 45; operational 31,in short term repair 8)

Berge Panther (quota 5; operational 3, short term repair 1)

Flak Pz IV (quota 8; operational 6, short term repair 2)

Comments: 2 absent officers, written request made out for 2 replacements, also 1 Doctor, and 10 men.

Short judgment by the Battalion Commander

Training: *The training position of all the companies is good due to improved running, especially since the time spent on infantry training. Above all, time has been spent on tank-fighting measures. Supply and repair services are now convenient. Instruction for technical service by specialists has been combined.*

Morale of the troops: *The troops are mission ready and particularly prepared and ready with endeavor to achieve the best results. Current perception and evaluation, they have taken on the responsibility for these unique weapons so far as their reliability in operation is secure.*

Special difficulties: *In the reports sent, technical difficulties encountered on gear change, steering, and on side transmission are the cause of the breakdowns. On the whole these are now possible for remedy by the Battalion.*

Judgment: *The Battalion is fully suitable for each attack and defense.*

Major and Battalion Commander.

33.3 Waiting to counter the Allied attack

For the first half of March, s.Pz.Jg.Abt. 653 had been preparing for combat in Haguenau Forest, all vehicles and supplies being heavily camouflaged. Dozens of Allied aircraft had made sorties over the area and any exposed vehicle had been attacked. Limited fire was returned so as not to attract attention to the unit's positions. Radio traffic was also prohibited, all communication being by motorcycle dispatch riders.

Between 1 and 11 March, the Workshop Company had repaired 7 Jagdtigers with short-term breakdowns, which had been conveyed to the Haguenau Forest for operations.

By 15 March, 41 Jagdtigers were in the battle area. 1/653 (Haberland) with 13 Jagdtigers was near Biblisheim-Gunstett, just

Plate 255. A well camouflaged Jagdtiger in the Haguenau forest. Aerial reconnaissance by the Allied airforce detected the Jagdtiger Battalion resulting in numerous air attacks (Karlheinz Münch).

Plate 256. In the heart of the Haguenau Forest, a camouflaged Jagdtiger takes on fuel ready for the forthcoming operations (Karlheinz Münch).

Plate 257. Men of the s.Pz.Jg.Abt 653 prepare a 'log bunker' (Karlheinz Münch).

Plate 258. Aerial photograph of the front line and the military camp in Haguenau. The Moder River was not very wide at this point and did not present the Allies with a difficult river crossing (U.S. Army).

off the road. The 2/653 (Wiesenfarth) was behind Reichshoffen with 14 Jagdtigers and 3/653 (Kretschmer) were near Gundershoffen.

Over a three week period the command carried out extensive reconnaissance and mapping of the area. Plans were made and positions prepared, for attacks at any point along the Moder front. They could move into pre-prepared positions to counter an attack at any point.

All the bridges over the Moder front had been destroyed since OPERATION NORDWIND had been suspended on 25 January. The infantry in this sector on the front were the 905th and 257th Volksgrenedier Division. They had had time to get well dug-in. All the snow had gone, and very damp conditions now prevailed. 653 had almost 1000 men themselves in the area.

There was no shortage of supplies of food, ammunition or even fuel. OKH had seen to this, as the Fuhrer himself had shown great interest in the Jagdtiger Battalion.

The troops were well rested and ready for action. The last Jagdtigers arrived in the area on the night 10/11 March. All were conveyed on flat rail cars with battle-tracks fitted. They were unloaded and driven straight into the woods. Fortunately, and due to the crew's skill, there were no Jagdtigers lost during the move south to this area.

Even with the strictest regime of diligent camouflage the Jagdtigers were subjected to air attack, their track marks acted as useful guides to the Allied Airmen.

An attack over the Moder was expected at any time. The moral amongst the troops was high. Many were veterans from the Eastern Front and the Italian campaign and were well used to combat in the Ferdinand tank destroyers. They rated the Jagdtiger as better all round as far as protection was concerned, only its mechanical reliability was suspect. Additionally they had to be very careful with the alignment of the gun optics, the gun support-frames were released at the very last minute.

The Panzer Jägers wanted to prove themselves with this new weapon.

There had been some limited twilight firing, in early March at troops over the river, tanks and lorries were hit at ranges up to 5km.

On one occasion Fw Herman Luft, a gunner for 1/653, scored his 25-tank kill when he shot up an Allied column. *He was awarded the Tank Fighting Medal in Silver, on 6 March 1945. Major Grillinberger presented it.*

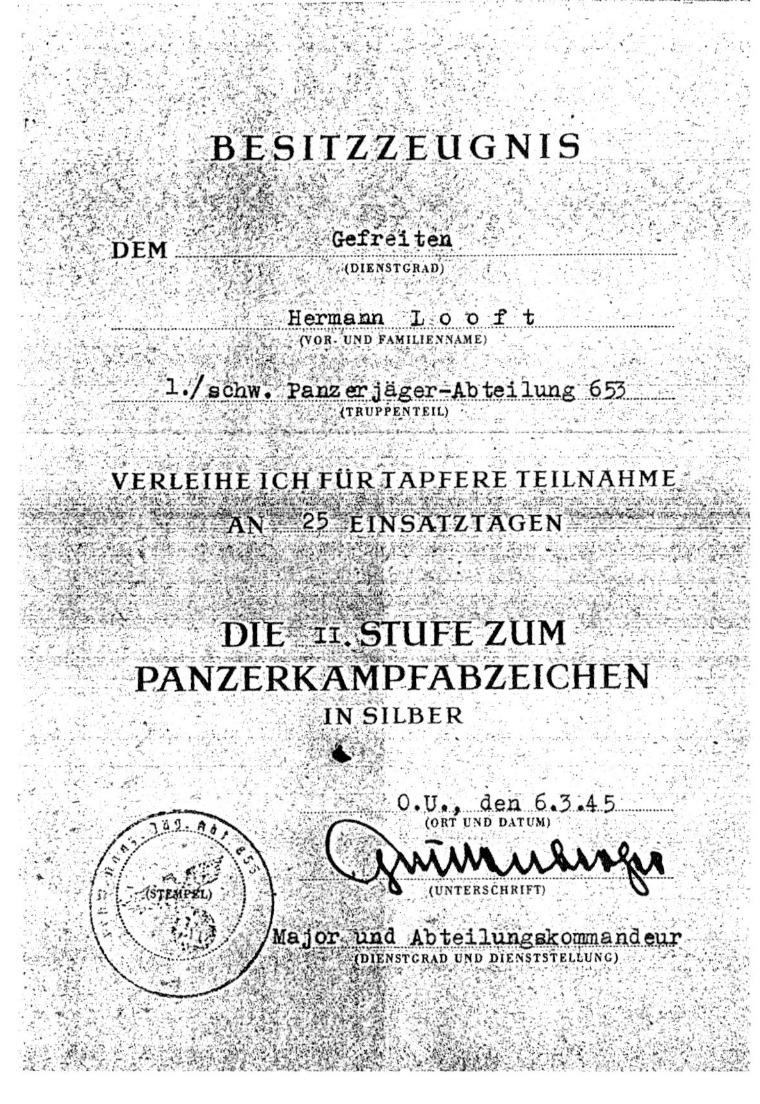

BESITZZEUGNIS

DEM Gefreiten
(DIENSTGRAD)

Hermann L o o f t
(VOR- UND FAMILIENNAME)

1./schw. Panzerjäger-Abteilung 653
(TRUPPENTEIL)

VERLEIHE ICH FÜR TAPFERE TEILNAHME
AN 25 EINSATZTAGEN

DIE II. STUFE ZUM
PANZERKAMPFABZEICHEN
IN SILBER

O.U., den 6.3.45
(ORT UND DATUM)

(STEMPEL)

(UNTERSCHRIFT)

Major und Abteilungskommandeur
(DIENSTGRAD UND DIENSTSTELLUNG)

Because of the Jagdtigers mechanical unreliability, road marches had been limited to a minimum and careful driving, particularly in third gear, was to be observed at all times. Most of the Jagdtigers still had the suspect steering units. Time had not been available to change them all. Neutral turns had to be avoided at all costs.

The works unit remained at Bellheim to await damaged or broken vehicles. Hauptmann Ulbricht of Support Company had 4 operational Bergepanthers in the forest ready to tow Jagdtigers to rail evacuation points and then on for repair.

As the Jagdtiger was one of Hitler's secret weapons (war winners), he expected much of the unit. This put great pressure on Major Grillinberger and almost weekly, reports were being sent to Insp. d. Pz. Tr. (Major General Thomale).

The fact that the command of 653 had been criticized in the specialist investigation report, in mid-January, this had put both Haberland and Grillinberger's positions at risk: they had had a severe dressing down by OKH.

Grillinberger's orders were to deploy with Army Group G in the Wissembourg depression behind the left flank of the First army command. They were to destroy any Allied armor that tried to penetrate the plain behind the Haguenau Forest and prevent any armored attack from breaking out of the bridgehead positions. They were also to provide artillery support for the infantry units.

Strict orders were issued to the three company commanders, under no circumstances were any undamaged Jagdtigers to be retrieved intact by the Allies.

All possible effort had to be made to retrieve damaged Jagdtigers, should this prove to be impossible then specific instructions for the use of the two demolition charges had been issued but only at the last minute.

33.4. Report dated 11 March 1945.

Telegram: To OKH/Gen. Insp. d. Pz. Truppen
From the s.Pz.Jg.Abt. 653,

To OKH on 11/03/45 stated; Position on 11/03/45 - 41 Jagdtigers (38 operational, 3 require repair), 4 Flak Pz IV (3.7cm), 4 Flak Pz 1V (2cm), 1 Flak Pz V (2cm v), 4 Bergepanther, 14m SPW.

Plate 259. A well-camouflaged Jagdtiger opens fire on the Allied positions. The huge amount of smoke generated by the 12.8cm ammunition is evident. 14 March 1945 (Karlheinz Münch).

The unit had reached its maximum strength!

33.5 The s.Pz.Jg.Abt.653 go into action

From their positions near Surbourg, the Jagdtigers made an artillery attack (14km range) on an American observation post in Haguenau on the South bank of the Moder River, it was reported hit several times.

Early in the morning (0700 hours), on 15 March, the troops of 1/653 and 2/653 were alerted to an impending Allied attack by the thunder of a heavy artillery barrage, 12-15km to the southeast of their positions.

The target of the attack was the village of Oberhoffen and the military camp between there and Haguenau. Two thousand shells preceded the Allied attack, which reduced Oberhoffen to a pile of rubble. The German ground troops were well dug in and sat it out until the artillery barrage stopped. French overseas Algerian, Moroccan and Tunisian troops crossed the river and into Oberhoffen. A bridgehead was quickly established after some bitter hand-to hand fighting. The fighting continued throughout the morning, while a bridge was being built to carry the heavy equipment across the Moder River. German artillery pounded the south bank in an effort to prevent the bridge from being completed.

By 0830 hours, information about the attack had been passed to 653 who were ready with engines warm for the drive to the battle area. Wiesenfarth with his 14 well-camouflaged Jagdtigers were in the Reichshoffen-Schirlenhoff-Eberbach area. The company guarded the roads from Zinswiller.

Haberland with his 13 Jagdtigers had been deployed into positions north of the edge of the forest near Gunstett-Biblisheim, from the slight ridge they had a commanding view of the plain towards Mertzwiller.

A Bailey bridge was established over the Moder by late afternoon. The French put Sherman tanks across the river into Oberhoffen. In the early evening, a tank assault was attempted and stopped by artillery.

By the evening, the Allied forces were still pinned down in the gas works on the edge of the village. As darkness fell, Shermans attempted to cross the railway to support the troops in the gas works. This would then allow a heavy attack to be mounted on the military camp.

All through the night, the battle raged around Oberhoffen and the military camp. Mortar and artillery fire was directed at the defenders. Neither side was able to gain ground from the other. Infantry units of both forces sustained very heavy casualties.

By the morning of the 16 March, 1/653 had 7 operational Jagdtigers. The recovery unit with 2 Berge Panthers and 2 18t half-tracks prepared to withdraw damaged vehicles 8km back to the railway station in Soultz as darkness started to fall.

Grillinberger sent a Hptm Ulbricht to arrange the trains to evacuate the damaged Jagdtigers to the works area at Bellheim. The journey and loading operations were to take place under the cover of darkness.

The 2/653 had Jagdtigers out of action. These were being towed by two other Berge Panthers 12km along the forest roads through Gunstett Surbourg to woods near the station at Soultz for evacuation.

The battle for the military camp and Schirrhein raged all through the daylight hours. The frustrated attackers resorted to massed artillery and dozens of fighter-bomber sorties to soften the German defense.

Plate 260. Jagdtiger tactical number 101, Oblt Haberland, on the move in the Haguenau area at twilight to avoid air attack (Karlheinz Münch).

Plate 262. Side view of 102 chassis number either 305006, 305007 or 305008 (Tank Museum).

Throughout the 16 March the Algerian infantry gradually gained control of Oberhoffen and Schirrhein, but no supporting armor was able to get through. By 1900 hours, the battle for the military camp was over. Allied infantrymen had got into the forest north of the camp and started to move north through the forest.

By midnight, two trains were loaded with 10 Jagdtigers. These were dispatched north for Bellheim. A Kampfgruppe of seven Jagdtigers remained led by Oblt Haberland. They took up position on the edge of the forest near Biblisheim to protect the loading operation. The Kampfgruppe in conjunction with 3/653 held off a U.S. evening attack, which was pushing north from Laubach for Gunstett thus protecting the withdrawal from the area.

That night, in the area of Schirrhein, German forces launched one last counter attack. The 257th Volksgrenidiers hit the Algerian infantry and struggled for three hours to win back control of the village. However, the attack failed, the infantry continued to fight for the forest. The eastern flank of the Moder Front had been penetrated.

After the trains had left, the Kampfgruppe withdrew. One Jagdtiger, No 102, broke down and was blown up near Rittershoffen. The others were carefully driven or towed north through Surbourg and Hunspach to Wissembourg, a distance of 15km. One Jagdtiger from 2/653 broke down just before Wissembourg and had to be blown up The American attack did not resume in this sector until daybreak on the 17 March.

The two trains were loaded by 2100 hours and dispatched north. The Allies had broken through the forest by midnight as the Kampfgruppe drove down near Seltz.

Plate 261. Jagdtiger tactical number 102 after its thorough destruction – both demolition charges were used to great effect (Tank Museum).

Plate 263. Oblt Kretschmer with his crew stand in front of a Henschel Jagdtiger (Karlheinz Münch).

Plate 265. Front view of tactical number 314, Chassis number 305012. It was painted in the ambush camouflage scheme (U.S. Army).

After the collapse of the Haguenau/Moder front, the positions along the Siegfried Line were occupied as the next defensive barrier in this sector.

The report sent to OKH by Grillinberger stated:

34 Jagdtigers: 18 operational (13 road march, 5 rail to new area) 16 needing repair, all on rail transport.

33.6 3/s.Pz.Jg.Abt.653 counter American attack

In mid March, the Allied strategy was to mount a two-prolonged attack on the Moder front; a frontal assault by French colonial forces near Haguenau with a simultaneous American attack at Ingwiller. Its objective was to surround the German forces in the Haguenau Forest.

The American plan was to push northeast to Wissembourg, then due east on the flood plain between the forests of the Haguenau and Bienwald. After mopping up operations, the Siegfried Line could then be assaulted.

The American offensive started on the morning of 14 March with artillery and fighter bomber strikes being directed at the German position. A bridgehead was quickly achieved over the Moder River.

The 3/653 under the command of Oblt Kretschmer was positioned in the woods near Gundershoffen with fourteen Jagdtigers, a mere 12km north east of the U.S. bridgehead.

The Americans soon became well established in the southern part of the Foret d' Ingwiller and were forming for a concentrated attack to the northeast to strike towards Wissembourg.

On the night 14/15 March the 3/653 were camouflaged in a wood in the Laubach-Griesbach area. The Americans were in Mertzwiller and had halted for the night. The Jagdtigers had a long field of fire, flares were fired and the column of U.S. tanks could be seen stretched out along the road. The experienced gunners of 3/653 aimed well and let loose their 12.8cm cannons. At 3 to 4km ranges the results were extraordinary, even better than the 8.8cm cannons of the Ferdinands, with which they had

Plate 264. An impressive side view of Jagdtiger No 314. It lost its transmission near Morsbronn, 16 March 1945 (U.S. National Archives).

Plate 266. Front view of 314. Jagdtiger tactical number 332 can be seen in the background (Tank Museum).

Plate 267. Left side view of Jagdtiger 332, after being hit by a rocket, which damaged its running gear. It was blown up by its crew in the early hours 17 March (U.S. National Archives).

previously fought. The tank column was shot up at the front and rear, the rest tried to get off the road, they eventually returned fire but to little or no effect. After 10 minutes a heavy barrage of artillery shells were being directed at the woods. Kretschmer withdrew the Jagdtigers without loss; he was a very experienced officer and had been company commander with 3/653 on the Eastern Front in the fighting around the Ternopol area.

The following morning 15 March, After artillery and fighter-bomber strikes the U.S. attacked resumed, striking across country towards Morsbronn. The Americans occupied the ridge at Forstheim and tried to move forwards. They were hit by very accurate 128mm fire and suffered high losses.

A senior officer ordered the 3/653 to attack the ridge in the afternoon. The Panzer Jägers followed orders and broke cover to assault the position. After 2-3 kilometers the Jagdtigers were attacked by P-47 Thunderbolts which attacked 3/653 with air to ground rockets. Jagdtiger No 332 was hit in the right side, Jagdtiger No 314 lost its transmission and ran up a mud bank at the side of

Plate 268. Right side view of 332 (Tank Museum).

Plate 270. Further right side view of No 332. (Tank Museum).

Plate 269. Front view of 332, it was painted in the three-color ambush camouflage scheme (Tank Museum).

the road. Three other Jagdtigers were also hit along the 4km route between wooded areas, one of these Jagdtiger No. 302 had been traveling at speed and survived multiple near misses with only damaged skirt armor. The flak vehicles returned fire and one aircraft was hit.

Interestingly the Allies claimed 4 Jagdtigers destroyed by the air attack! Veterans claim the four damaged Jagdtigers as being self-destructed because recovery was not possible on the 17 March. The Keystone Press Agency pictures state "Jagdtigers destroyed by air attack on 16 March", they feature Jagdtiger 314 and 332 which were no longer burning!

On 16 March, the Americans who were unable to move tanks in the area directed a vicious artillery attack at 3/653 and two Jagdtigers were damaged, one was destroyed No 301. The company commander Kretschmer was wounded, through concussion he was unable to continue command. His driver, Uffz Appel with other crew members got him out of the vehicle then used the demolition kit.

The remaining seven Jagdtigers managed to get into the woods near Woerth. Command was taken over by the quick thinking Battalion Ordnance officer, Lt. Herman Knack. He restored the situation and put the Jagdtigers into a firing position on the forest road between Woerth and Merkwiller. They engaged pursuing American forces late in the afternoon.

Knack was informed by radio that the broken down Jagdtigers of 1/653 and 2/653 were to be withdrawn from the Haguenau Forest that night. Many of their Jagdtigers were out of action.

Plate 271. French soldiers inspect Jagdtiger tactical number 301, chassis number 305010. It had thrown a suspension unit, and an artillery round dislodged its gun mantle (Leclerce).

The Americans had to be prevented from cutting the Haguenau - Soultz railway line, the 653's rail evacuation route.

The 3/653 kept the Americans at bay in Morsbronn but with the temporary suspension of the attack northward, they tried to push east through Gunsett, only to be stopped by crossfire from 7 Jagdtigers belonging to 1/653.

After a successful evacuation of the majority of 1/653, on the 17th, Knack was ordered to withdraw his Jagdtigers along the forest roads of the Hockwald through Lampertsloch, Drachenbroon to Wissembourg, through which they traveled at night. The withdrawal avoided further air attack, with very careful driving being

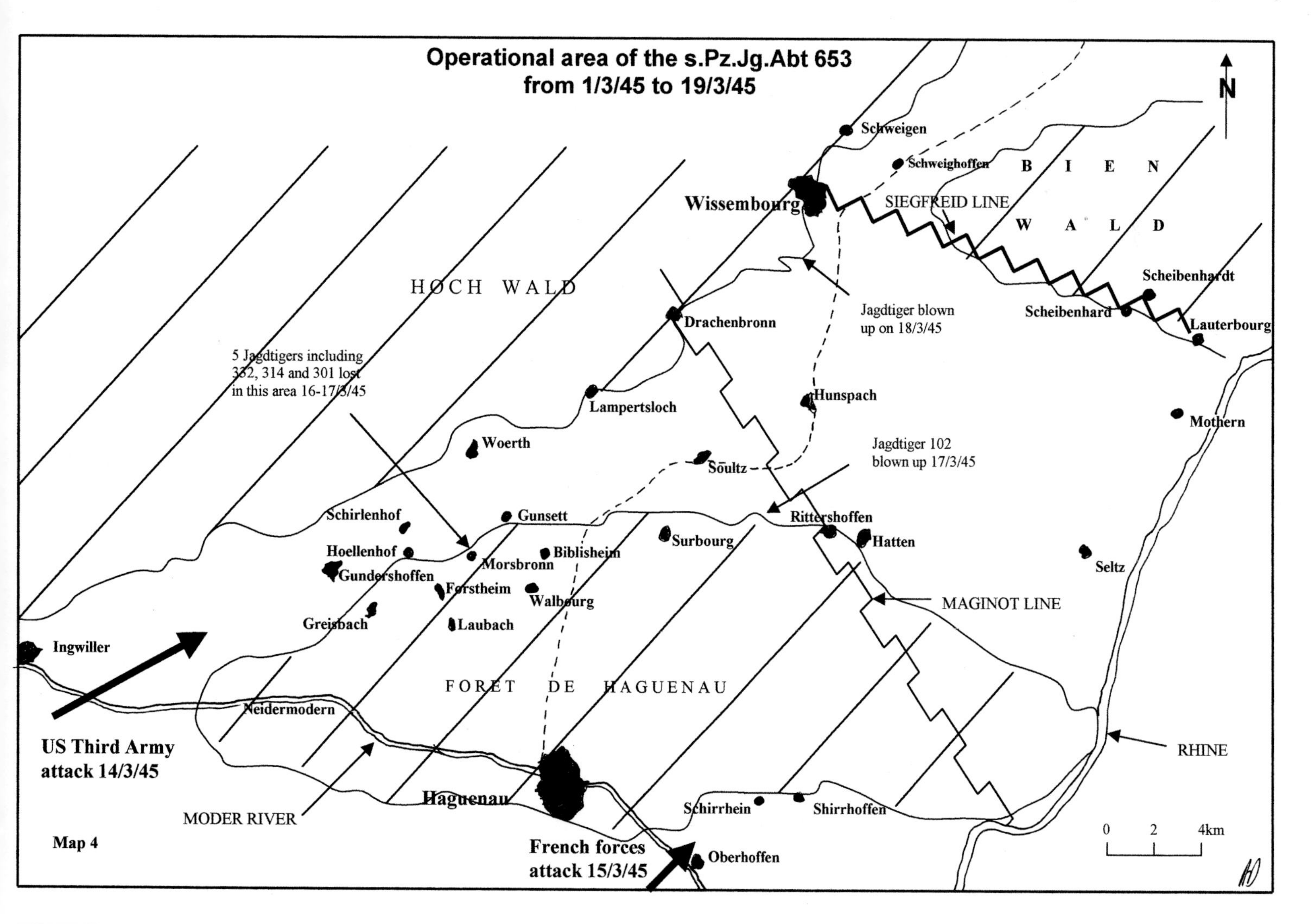

observed by these more experienced troops. In some instances, operational ones towed the broken down Jagdtigers!

By the night of the 17/18 March, the 3/653 was in woods near Schweigenhofen and behind the Siegfried Line, Knack reported the Jagdtiger's losses to Major Grillinberger. 3/653 was ordered to assemble with the rest of the Battalion at Bellheim. One Jagdtiger, which had been towing another, broke down on a hill.

33.7 Last actions west of the Rhine

After the withdrawal from Haguenau Forest, by 18 March, thirty-four Jagdtigers had been withdrawn behind the Siegfried Line. This was a First World War-type of defense system made up of concrete bunkers and dragon's teeth. It was heavily overgrown and made an excellent barrier against the Allied forces in the south.

Sixteen Jagdtigers of 1/653 and 2/653 needed repairs as well as a further four belonging to 3/653. The Battalion was making their way to Bellheim to the works assembly area. Twenty-one Jagdtigers were on trains with thirteen either driving or being towed back. There were numerous air attacks directed at the Jagdtigers.

By 21 March, Hitler (because of the impending collapse of the forces west of the Rhine) issued the orders that:

Fernschreibstelle 348

Fernschreibname Laufende Nr.

Angenommen / Aufgenommen
Datum: 21/3 194
um: 1910 Uhr
von: HOLG
durch: [illegible]

Befördert:
Datum: 194
um: Uhr
an:
durch:
Rolle:

Gen. Insp. d. Pz. Tr.
Eing.: 21 MRZ 1945
Nr. 4452/45

Vermerke:

++--FRR-- WRKF 1729/32 21.3. (1800)== -- QEM ---
NACHR.: GENERALINSPEKTUER DER PANZERTRUPPEN==
GLTD: OB WEST==
OB WEST/ OQU.-
GENST D H/ GEN QU ZEPPELIN.=
NACHR.: GEN INSP D PZ TR.=

Bestimmungsort

-- GEHEIM--.-
DER FUEHRER HAT BEFOHLEN, DAZ DIE BETRIEBSSTOFFVERSORGUNG
DER JAGDTIGER ABT 653 DURCH GESONDERTE ZUFUEHRUNG UNTER
ALLEN UMSTAENDEN SICHERGESTELLT WIRD. OB WEST MELDET DAS
VERANLASZTE UND DIE HOEHE DER BEVORRATUNG.==
DER CHEF DES OBERKOMMANDOS DER WEHRMACHT GEZ. KEITEL.
GEN. FELDM .++

Plate 272 Jagdtiger hit by artillery in Iggelsheim (Karlheinz Münch).

The s.Pz.Jg.Abt 653 be, at all costs, saved for future actions east of the Rhine before the bridges were blown up.

By this day the Siegfried line was withstanding two assaults by Allied forces. General Monsabert with his French Colonial forces had been attacking at Schiebenhardt with little success and, on 22 March, after a 48-hour barrage, General Brooks with the U.S. forces struck at Wissembourg. At the very last minute a Jagdtiger which had broken down near Schweigen an Weintor was blown up to prevent capture, when the Americans broke through.

The greatest threat to 653 was the American attack pushing south from Ludwigshafen. On the night of 21/22 March, as the first trains were evacuating eighteen Jagdtigers over the Rhine at Germersheim, Grillinberger deployed Kampfgruppen from the three fighting companies, to hold the line Neustadt-Böhl-Schifferstadt and stop the U.S. attacks which threatened the Battalion's evacuation over the Rhine.

The 1/653 was in Schifferstadt with 3 Jagdtigers under the command of Oblt Haberland. One American attack was stopped in the early hours of 22 March, a second attack, which was forming in the late afternoon, was also shot up at long range, the Americans withdrew.

The Jagdtigers held the position inspite of air and artillery attacks, one of which wounded the company commander, he was taken to hospital. At least four U.S. tanks and armored cars were destroyed.

The three Jagdtigers were evacuated over the Rhine on the night of 23/24 March at Speyer.

A second Kampfgruppe with 3 Jagdtigers from 2/653 commanded by Oblt Wiesenfarth were deployed in the town of Böhl. In the morning of March 22, a strong American tank attack set off from Hochdorf towards their positions. The Jagdtigers knocked out 9 tanks and 2 armored cars. The Americans withdrew and directed artillery at the Jagdtigers.

A fierce battle erupted led by American infantry. Its crew blew up one Jagdtiger as it was being over run by infantry in Böhl. A second withdrew to a new firing position and was hit by artillery at Iggelheim, it burned out and the third was able to withdraw across the Rhine with the 1/653 Kampfgruppe.

Lt. Kasper Geoggler, in Jagdtiger No 331 commanded the third Kampfgruppe from 3/653. He placed the three Jagdtigers into a good position north of Neustadt with a commanding view of the approach roads to the town.

Geoggler had nerves of steel, and was very keen to prove himself, he had been awarded the Cross-of Gold, on the 10 May 1943 whilst fighting on the Eastern Front, and already he had several kills to his credit with his Jagdtiger.

From camouflage positions, the three Jagdtigers engaged a U.S. tank column; the first and last vehicles were shot up followed by the rest. The Shermans and M10 tank destroyers returned fire. Two Jagdtigers, Geoggler's and another, No 323 were hit ten times between them, they withdrew into Neustadt. Twenty-five U.S. tanks were claimed destroyed, none of the Jagdtiger crews suffered any serious injuries, and the thick sloping-armor had done its job.

In Neustadt, No 323 broke down while trying to maneuver into a firing position in a garden. On the road opposite, No 331 was placed outside House number 70, Landauer Strasse, to fire

Plate 273. Photograph of Jagdtigers tactical number 331 and tactical number 323, in Neustadt an der Weinstrasse (U.S. National Archives).

Plate 274. Jagdtiger tactical number 331 chassis number 305020, after abandonment, on 23 March 1945 (U.S. National Archives).

Plate 275. Side view showing how short No 331's gun appears after it was sabotaged into the full recoil position (Aberdeen Proving Ground Museum).

Plate 276. Further view of No 331 undergoing examination in Neustadt (Aberdeen Proving Ground Museum).

up the road, when its last round went into the chamber, the recoil cylinder was drained and the gun fired, this jammed it in the full recoil position, and unusable to the Americans.

Geoggler's men left with the half-tracks and the third Jagdtiger No 234, they drove back to Zeiskam where No 234 was again in action at the railway level crossing, a shot broke its right track, after which it was abandoned and blown up. Kampfgruppe Geoggler withdrew through Iggelsheim-Dudenhofen and crossed the Rhine at Germersheim with the rest of the Battalion, their defense task successfully completed.

33.8 Assessment of s.Pz.Jg.Abt.653 by Oberstlt. (Ing) Meyer - 23.3.45

ObW had sent Meyer to report on the movement of s.Pz.Jg.Abt 653. His telegram report was as follows:

Plate 278. View of No 323 under new management – the French moved it to the Festival platz (Delta Publications).

Plate 277. View of the impact damage sustained by Jagdtiger 323. It was photographed in the place where it was abandoned in Neustadt. (Aberdeen Proving Ground Museum).

Plate 279. Rear view of No 323 (Delta Publications).

Plate 280. Right side view of No 323, in the Festival platz, in Neustadt Weinstrasse. It was scrapped in this position in 1948 (U.S. Army).

Plate 281. Front view of No 323, without its gun-clamp. Chassis number 305022 is just visible on the original photograph (U.S. Army).

1. Before 18 March 1945, 653 had 41 Jagdtigers and were situated in the perimeter of the Western front. 7 Jagdtigers had to be blown-up because of damage and towing was impossible.

a. 18 Jagdtigers were still operational, 5 on rail-transport, 13 on road-march to new battle area.

b. 16 Jagdtigers in need of repair were on rail-transport to the maintenance area of Bellheim near Germersheim.

2. The Jagdtigers in the area of Bellheim, along with the Bergepanthers, 4 flak tanks and heavy half-tracks were loaded and transported on 21/22 March and sent to Graben-Neudorf, (these had been taken across the Rhine to a safer area).

3. There had been no setbacks during the transportation of the 18 Jagdtigers. Oberstlt Meyer observed these tanks in action. Sufficient fuel was available to fulfill action requirements.

33.9 Further report - 23.3.45 (Telegram)

Ref. the s.Pz.Jg.Abt. 653 Jagdtigers:

Tank position on 22.3.45 at 2000 hours. Total Jagdtigers available 31, 2-3 fully operational,(1) *9 vehicles on rail transport for evacuation over the Rhine.*

Throughout the nights 22/23 March, 2 Jagdtigers lost through explosion - reason not known.(2)

Major Grillinberger needs all the Jagdtigers urgently for 2 weeks for maintenance reasons.

The unit was fighting in the Neustadt - Landau and Iggelheim areas on 22 March 1945.

By 25 March, there was no further resistance to the Allies on the West Bank of the Rhine.

33.10. Fighting east of the Rhine

After the successful evacuation of the Battalion across the Rhine, they were now busy setting up a new camp in the wooded

Plate 282. Side view of No 234, as it is being examined by curious Allied infantry (U.S. Army).

Plate 283. Jagdtiger No 234 was stopped by a shot, which broke its right track. The suspension lowered after it burned out (Delta Publications).

Plate 284. Jagdtiger of the 2/s.Pz.Jg.Abt 653 in the area of Bruchsal, on March 27 1945 (Karlheinz Münch).

Plate 285. Jagdtiger tactical number 115 after its destruction in a field outside Schwetzingen (Karlheinz Münch).

area near Neudorf. This is an area close to the Karlsruhe - Mannheim rail line and the Landau - Bruschal line and is heavily covered by forest with numerous small rivers and streams that run northwestward into the Rhine. The Works Company were positioned 4km northeast of Bruchsal at Unteröwisheim.

On 24 March, the Rail Bridge at Germersheim was blown up, the 653 battalion were busy trying to get back to operational readiness. At this time the panzer positions was:

1/653 had 12 remaining Jagdtigers
2/653 had 10 remaining Jagdtigers
3/653 had 6 remaining Jagdtigers

Because of the excessive number of broken-down vehicles, Oblt Schulte the works company commander ordered an immediate assessment to establish which were the easiest vehicles to repair. He had applied to Major Grillinberger for a two-week standdown for the Battalion to gain time for effective repairs to the Jagdtigers; the Major had passed this on to the Insp Gen d Pz. Troops.

On the 25 March, action was taken when Grillinberger was summoned to OKW and demoted to Lieutenant "for tactical blunders". He was blamed for the rapid collapse of the Moder Front, Grillinberger had never been popular with his troops.

Command was taken over by Major Rolf Fromme on the same day. He was the former Battalion commander of the Tiger unit s.Pz.Abt. 503 from the 20 February 1944 to the 9 December 1944. He was a very experienced Panzer commander with combat experience in Russia, Normandy and lastly Budapest with Tiger II's. He was used to leading from the front; unfortunately, he was not fully familiar with limited traverse vehicles, and he had no experience of the Jagdtiger itself.

Fromme inherited a force of twenty-eight Jagdtigers, nineteen of which were out of action and troops with low confidence in the Jagdtiger's mechanical reliability, although they were very satisfied with its armor and gun.

The Works Company had barely three days to work on a limited number of Jagdtigers. On 26 March, the Americans crossed the Rhine at the Germersheim just south of Worms, a mere 40km north of 653's position. They still had only nine Jagdtigers combat ready.

Fromme, who had little time to meet all his staff, was in possession of the only heavy armored unit capable of countering the U.S. attack.

On the evening of 28 March, Lt. Hans Knippenberg, platoon (zug) commander of 1/653 was presented with the Eiserne Kreuz I. Klasse (Iron Cross 1st Class).

IM NAMEN DES FÜHRERS
VERLEIHE ICH
DEM
Leutnant
Hans Knippenberg
1./schw.Pz.Jäger-Abt. 653

DAS
EISERNE KREUZ
1. KLASSE

Div.Gef.St., 28.3. 1945

Generalleutnant und Div.Kdr.
(DIENSTGRAD UND DIENSTSTELLUNG)

Plate 286. Front view of tactical number 131 (U.S. Army).

Plate 288. Right side view of 131 (Karlheinz Münch).

Rolf Fromme made the presentation. Knippenberg's award was based on his successes in the Rittershoffen area.

As daylight broke on the morning of 30 March, the town of Schwetzingen, was attacked by fighter/bombers, fifty-six civilians were killed. After the bombardment the U.S. ground forces started to drive south into the town, they came into Schwetzingen from the west. Three Jagdtigers had been ordered to stop the American advance and were moving into the town from the south.

One Jagdtiger No 115 commanded by Lt. Knippenberg got stuck in a field while trying to maneuver into a firing position and it was blown up.

Jagdtiger No 131 was driving down Heidelbergstrasse when a Sherman in Mannheimstrasse fired and hit it in the left side, jamming its left track, it swung out of control into the house of the Krebs family on the corner of the two streets. A second shot fired at a range of 100m by the Sherman hit the Jagdtigers side armor above the wheels. The tank burned out with the loss of the one crewman with another fatally wounded.

The third Jagdtiger, which was also a Henschel vehicle, quickly drove out of Schwetzingen in an easterly direction towards Eppelheim but when it reached the E5 Autobahn it threw a track whilst at full speed. The Americans seemed to be everywhere and as nothing could be done to retrieve the Jagdtiger or put it into a firing position, the demolition kit was used.

There was now no resistance to the U.S. advance as it proceeded through Schwetzingen into Hoftersheim and Heidelberg, which were also captured on 30 March. Almost all the other German troops in the area, were just giving in, white flags were everywhere.

With the American forces a mere 15km north of the Battalion Camp at Neudorf, Fromme gave the order to evacuate this position and convey the non-operational Jagdtigers eastwards to a safer area. The 653 had made a reconnaissance, and a wooded area near Leonberg 15km east of Stuttgart was chosen.

That night, the trains were used to evacuate the 10 Jagdtigers, the rail route being through Bruschal, Bretten, Muhlacker, Schwieberdgen and Stuttgart to the new Maintenance area. A total transfer of 70km south east of Neudorf. Bruschal fell on 31 March.

The next in line of defense for the crumbling German forces in this southern sector was from Altrip - Rhein - Seckenheim Ladenberg Schriesheim to Großsachsen.

The strength report Telegram sent by Major Harder to Gen. Insp. d. Pz. Tr. At 1910 hours on the 30 March stated that out of

Plate 287. Rear view of 131. Its mono pod is in the fully extended position, with the engine hatch open (Karlheinz Münch).

Plate 289. A rather 'heavy' sign post (ECPA).

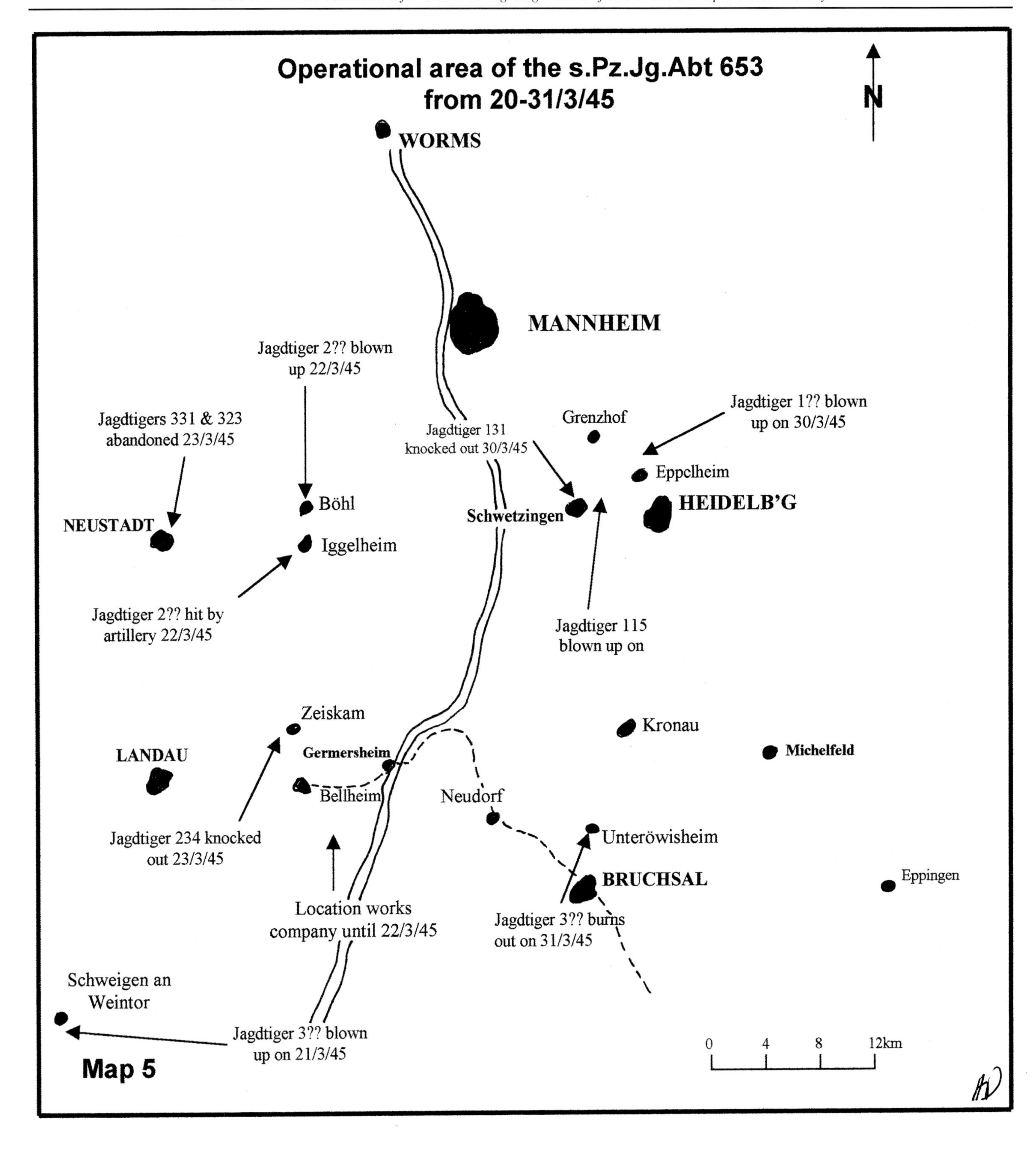
Operational area of the s.Pz.Jg.Abt 653
from 20-31/3/45
N
WORMS
MANNHEIM
Jagdtiger 2?? blown
up 22/3/45
Jagdtigers 331 & 323
abandoned 23/3/45
Jagdtiger 131
knocked out 30/3/45
Grenzhof
Jagdtiger 1?? blown
up on 30/3/45
Eppelheim
Böhl
Schwetzingen
HEIDELB'G
NEUSTADT
Iggelheim
Jagdtiger 2?? hit by
artillery 22/3/45
Jagdtiger 115
blown up on
Zeiskam
Kronau
LANDAU
Germersheim
Michelfeld
Bellheim
Neudorf
Unteröwisheim
Jagdtiger 234 knocked
out 23/3/45
BRUCHSAL
Eppingen
Location works
company until 22/3/45
Jagdtiger 3?? burns
out on 31/3/45
Schweigen an
Weintor
Jagdtiger 3?? blown
up on 21/3/45
0 4 8 12km
Map 5

twenty-eight Jagdtigers, six were operational (Schwetzingen Eppelheim area) with twelve needing minor repairs and ten long-term repairs.

Throughout a five-day period the Works Company had made one Jagdtiger ready for the troops to take over and a further one was now drive-able. There had been six new breakdowns since the 25 March.

With the relocation of the Works Company the rest of the battalion were making their way by road to the new assembly area between the Enz and Neckar Rivers to the northwest of Stuttgart.

33.11 Report dated 30 March 1945.

+ - - KR - - HGIX/ Qu Nr. 04639 30. 3. 1910 =
To Gen. Insp d. Pz. Tr. =
- - Secret - -
Panzer log of the s.Pz.Jg. Abt. 653: 28 Pz. Kpfw.-
Operational = 6 Pz. Kpfw.-
Short term repair = 12 Pz. Kpfw.-
Long term repair = 10 Pz. Kpfw.-
New breakdowns since the 25. 3. 45 are 6 Pz. Kpfw.-
By works platoon 1 Pz is repaired, one further panzer is ready to drive. -
On 29. 3. 45 early morning the Works Company and 10 long term broken down Panzers sent by rail to the area of Stuttgart. -
On 29. 3. 45 received from Ni-Werk 1 LKW. with 4 Porsche suspension units these met with in Bretten.-
1 LKW. With side drives from Cannstatt is missing. Search command underway. =
OBKDO. H. GR. G/ O. QU. / ROEM. FUENF/ PZ. TGB. NR. 221/45 G. V. 30. 3. 45 I. A. GEZ. HARDER, MAJOR +

33.12 Response 31/3/45

- - KR - - HOLG 1496 31/3 0135 =
To Gen. Insp. d. Pz. Tr. Herrn Generalleutnant Thomale. =

- - Secret - -
Major Harder V. Ob. West/ O. Qu. Report: -
Panzer log of the s.Pz.Jg. Abt. 653: 28 Pz. Kpfw.-
Operational = 6 Pz. Kpfw.-
Short term repair = 12 Pz. Kpfw.-
Long term repair = 10 Pz. Kpfw.-
New breakdowns since the 25. 3. 45 are 6 Pz. Kpfw.-
By works platoon 1 Pz is repaired, one further panzer is ready to drive. -
On 29. 3. 45 early morning the Works Company and 10 long term broken down Panzers sent by rail to the area of Stuttgart. -
On 29. 3. 45 received from Ni-Werk 1 LKW. with 4 Porsche suspension units these met with in Bretten.-
1 LKW. With side drives from Cannstatt is missing. Search command underway. =
Ob. West / O. Qu. 1 Nr. 2001/45 Secret
Affected. I. A. Coelle, Major I. G. +

33.13. Jagdtiger losses through out March 1945.

15 March 1945 *41 Jagdtigers (38 operational/3 need repair).*
Five Jagdtigers hit in air attack near Morsbronn

16 March 1945 Jagdtigers 301 lost to artillery fire in Morsbronn area.

17 March 1945 Jagdtiger 314 blown up in Morsbronn area
Jagdtiger 332 blown up in Morsbronn area
Jagdtiger 3?? blown up in Morsbronn area
Jagdtiger 3?? blown up in Morsbronn area
Jagdtiger 102 blown up near Rittershoffen

18 March 1945 Jagdtiger 2?? was blown up near Wissembourg
34 Jagdtigers (18 operational/16 need repair).

21 March 1945 Jagdtiger 3?? was blown up near Schweigen an Weintor

22 March 1945 Jagdtiger 2?? was hit by artillery near Iggelheim
Jagdtiger 2?? was blown up near Böhl
31 Jagdtigers (3 operational/rest in rail transport) at 2000hours.

23 March 1945 Jagdtiger 323 abandoned in Neustadt am der Weinstrasse
Jagdtiger 331 abandoned in Neustadt am der Weinstrasse
29 Jagdtigers (1 operational/rest in rail transport).
Jagdtiger 234 was blown up near Zeiskam.

25 March 1945 *28 Jagdtigers (11 operational/17 need repair).*

26 March 1945 *28 Jagdtigers (9 operational/19 need repair).*

29 March 1945 *28 Jagdtigers (8 operational/20 need repair).*

30 March 1945 Jagdtiger 115 blown up in Schwetzingen
Jagdtigers 131 lost at Schwetzingen
Jagdtiger 1?? Blown up near Eppelheim.

31 March 1945 *25 Jagdtigers (3 operational/22 need repair).*
Jagdtiger 3?? Burns out near Unteröwisheim

34

Schwere Panzerjäger Abteilung 653 April 1945

34.1 April 1945 combat synopsis

1 April The first day of April only 24 Jagdtigers were remaining, of these only 3 were operational in the Eppingen-Kirchardt area.

2 April Kampfgruppe withdraws through Muhlbach and relocated at Ochsenburg.

3 April 2 of 1/653 Jagdtigers (114 and 123) have to be blown up in Eppingen. There are 21 remaining Jagdtigers 1 operational. Kampfgruppe at Cleebronn.

5 April Kampfgruppe fighting the French in the Nordheim area one Jagdtiger is destroyed and a second breaks down, it is blown up. The group is relocated at Gemmrigheim.

6 April Withdrawal of Kampfgruppe to Besigheim, Jagdtiger 213 is blown up, they withdraw across the Neckar but Jagdtiger 214 has to be blown up in Lauffen because it cannot be taken across the river.

8 April Kampfgruppe now moves to Ludwigsburg.

9 April Battalion now down to 17 Jagdtigers, 10 had been repaired ready for action, 7 still needed major repair. Workshop Company loaded for transfer to Salzburg area.

14 April Throughout the transfer to Crailsheim 5 Jagdtigers break down. Strength report indicates 5 operational with 6 in short and 6 in long term repairs.

15 April Out of the 4 Bergepanthers, 1 operational with 2 needing short term repairs.

16 April OKH sends instructions to Major Fromme for a detachment from 653 to travel to Nibelungen Werk to take on 4 new Jagdtigers. The battle for Nürnberg starts.

17 April Spare Jagdtiger crews are ordered to Nibelungen Werk.

20 April 3 Jagdtigers lost in the Crailsheim-Nürnberg area, the City falls on Hitler's birthday.

24 April Battalion now scattered between Steinhaussen and Salzburg.

26 April Part of 653 has withdrawn to Walkertshoffen. Only 1 of the remaining 14 Jagdtigers is fully operational.

27 April Unit withdraws through Kissen near Augsburg.

28 April Unit moves to Schondorf am Ammersee. Augsburg occupied by U.S. forces.

29 April Moves through München/Geiselgesteig, Sauerlach. A detachment in Linz is ordered into action against advancing Soviet Forces.

30 April 653 have withdrawn to Bad Aibling. München occupied by U.S. forces.

34.2 Fighting around Eppingen

By the beginning of April the Allies had secured virtually the whole of the east bank of the Rhine, tank attacks had cut deep into the heart of Germany and the German Forces in the Ruhr area were surrounded.

In the south, Mannheim and Heidelbourg were already in Allied hands, their next objective was Heilbronn, to gain a bridge over the Neckar River. The vast majority of the German troops were now giving up without a fight.

On the first day of April the bulk of the 653 Battalion were involved with the move of base camps to the Leonberg area, a difficult move achieved with no time to spare.

The only operational Jagdtigers were Kampfgruppe Knippenberg who was ordered to Eppingen, a small village half way between Bretten and Heilbronn. Bretten was already in Allied hands.

The Kampfgruppe travel to Kirchardt to attack the American spearheads coming from Sinsheim.

Plate 290. The remains of two Jagdtigers from 1/653, Kampfgruppe Knippenberg. These were blown up just a few meters apart in the center of Eppingen (Karlheinz Münch).

Plate 292. Rear view of 114 (Karlheinz Münch).

Of the three Jagdtigers, Knippenberg's No 114 had lost its transmission in the village, Schlabs No 123 was also having problems, and the third Jagdtiger Kohns No 214, at this stage was running correctly.

With the mechanical problems being experienced by his Jagdtigers, Knippenberg ordered a towing operation to Eppingen. He hoped to be able to evacuate them from the railway station in Eppingen and convey them back to the Works Company for repair. This was not possible because the transport trains were still in Leonberg.

Knippenberg expected the Americans to attack Eppingen the following day.

The American advance had thus far encountered little resistance in the area until they approached Eppingen. Their lead tanks were again destroyed by accurate fire, three rounds from a single Jagdtiger No 214, and the U.S. tanks pulled back while artillery was brought to bear on the village. Knippenberg had no alternative other than evacuate the position. Jagdtiger No 114 and No 123 were blown up using their demolition kits, Jagdtiger No 214 and the support vehicles were driven east to Nordheim, and there were no casualties.

The two destroyed Jagdtigers and the three spent cartridge cases were later the subject of an Allied intelligence inspection[1] by Lt./Col. G.C. Reeves and Mr. Gouldthorp which was circulated on 10th May 1945. See following photos.

Plate 293. The blast from the demolition charges blew part of the gun out of 114, 3 April 1945 (Tank Museum).

Plate 291. (*left*) Front view of Jagdtiger tactical number 114 with the ambush camouflage scheme. The placement of the towing shackles indicate it as having been towed to this position by Jagdtiger 214 (Karlheinz Münch).

Plate 294. No 114 and 214 had been towing 123 into Eppingen, after it broke down on 2 April 1945. With no prospect of recovery they were blown up at the last minute (Tank Museum).

Plate 295. Further view of Jagdtiger No 114, the spent shell-cases bore testament to the fire fight (Tank Museum).

Plate 296. Rear view of Jagdtiger 114 after Hans Knippenberg blew it up in Eppingen, on 3 April 1945 (Tank Museum).

Plate 297. Further view of Jagdtiger 114 (Tank Museum).

Plate 298. A clear frontal view of 123 – the first two numbers of the chassis number can clearly be seen. It is interesting and disappointing that the chassis numbers were not recorded on the intelligence report (Karlheinz Münch).

34.3 Report dated 3 April 1945

Radio-signal: To OKW/Gen. Insp. d. Pz. Truppen
From s.Pz.Jg.Abt. 653,

Total 23 Jagdtigers, 1 operational, 11 need short term repair, and 11 need long term repair.
There have been 5 new breakdowns since 1/04/45.
5 total losses since 30/3/45.
Completed transport of company, not works company.
Transfer has proved difficult.
Last spare parts received on 30/03/45 at 1800 hours in Bretten.

Plate 300. Front view of No 123, after it had been blown up by its crew (Tank Museum).

34.4 Road block north of Stuttgart

The Workshop Company from its new location near Leonberg and other maintenance personnel worked flat-out on the easiest Jagdtigers to repair. Over 5 days 6 Jagdtigers were repaired. One Kampfgruppe entered Oschenburg on 2 April 1945, but there was no fighting.

After the fighting in Eppingen on 3 April, one Jagdtiger No 214 was able to withdraw and drove east to Nordheim. The crews from No 114 and No 123 Jagdtigers were evacuated with this vehicle.

At this stage the American attack pushed eastwards to try to gain a crossing of the Neckar River and capture Heilbronn. Fi-

Plate 299. Left side view of Jagdtiger 123 in Eppingen (Tank Museum).

Plate 301. Front view showing the ambush color scheme of 123 (Tank Museum).

Plate 302. The remains of Jagdtiger chassis number 305014 after self-destruction in April 1945. (U.S. Army).

nally, because the Germans had successfully destroyed the bridges on the 5 April, a French attack was diverted southwards in the direction of Stuttgart.

A further Kampfgruppe withdrew south through Oschenburg and then east along the road through Michelbach, Guglingen and then up hill near Cleebronn where the battalion command post was situated, which was reached before daylight on 3 April.

Two days later carrying heavy camouflage, they attacked a French tank column on the road near Nordheim. In the firefight one Jagdtiger was destroyed along with a number of French tanks. A second Jagdtiger from 1/653 broke down and was blown up south of Nordheim. They then withdrew further to the south to cross the Neckar at Gemmrigheim a third Jagdtiger No. 214 broke down and was blown up in Lauffen.

A Kampfgruppe moved into Besigheim on 5 April, and one Jagdtiger from 1/653 burned out due to an engine fire. The day after Jagdtiger 213 was blown up in the town because it could not cross the Neckar River. The troops moved into the Ludwigsburg area on 7/8 April.

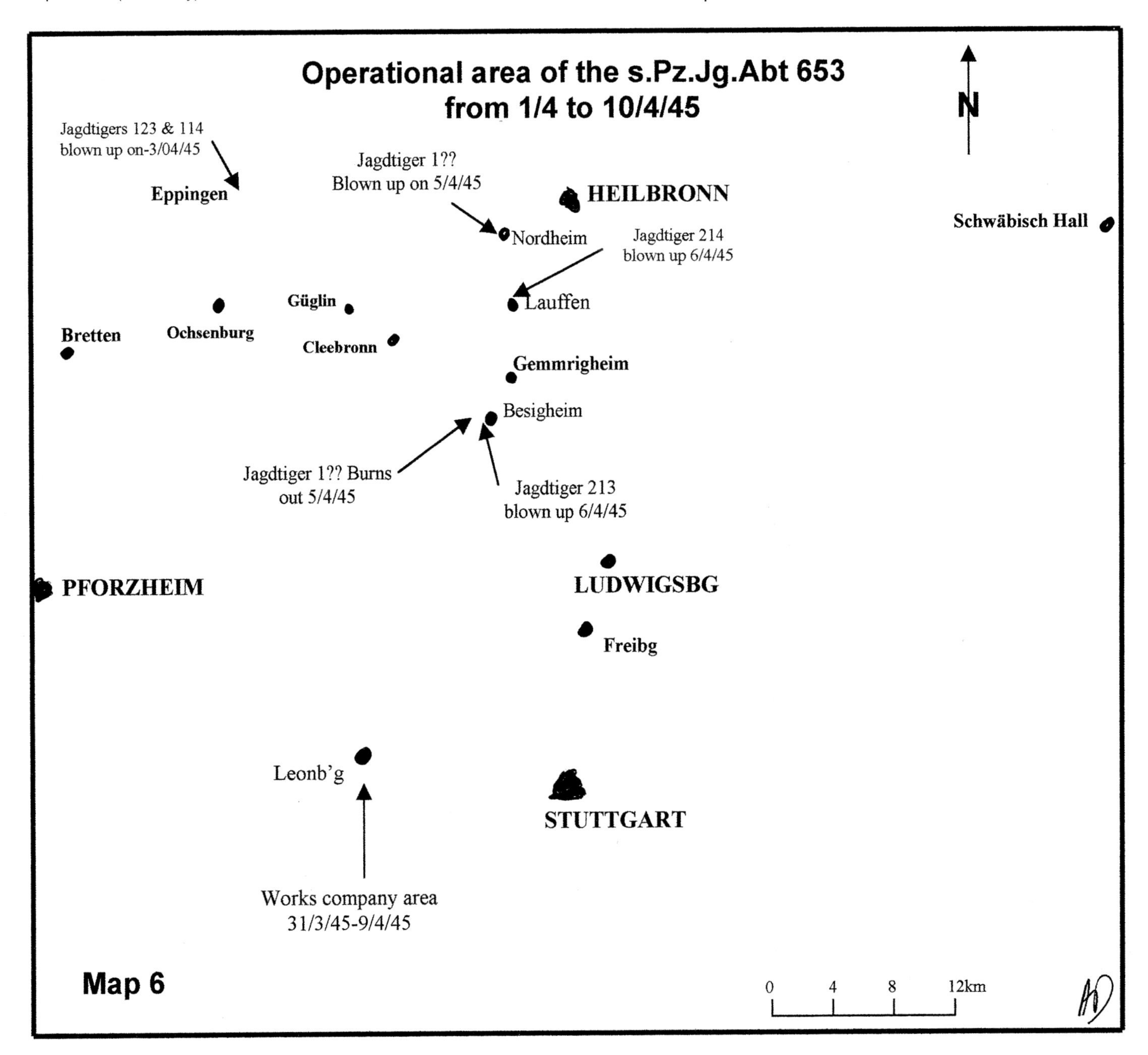

Another group was in action near Schwäbisch Hall on 7/8 April, they destroyed several American tanks, and it is thought that they lost one Jagdtiger in the area.

The Algerian and French forces in this sector were vastly superior and there was no alternative course of action other than to withdraw the German forces south of the Enz River. The Enz runs east through Pforzheim and joins the Neckar at Besigheim 15km north of Stuttgart.

Major Fromme had put top priority onto the repairing the Jagdtigers. Combat troops, railway engineers, as well as the works unit troops were all tasked at round the clock repairs. Railway cranes were also used in addition to their own equipment. On the 9th April, 653 were down to 17 Jagdtigers, 10 operational and seven in need of major repair (see later).

Faced with a river barrier on the Southern Front, the Allies concentrated on an eastern attack towards Crailsheim with the objective being to get behind Heilbronn. The rest of 653 were ordered to move to Crailsheim, on 10 April. That day over 100 German aircraft flew sorties over Heilbronn and Crailsheim attacking the American ground forces. Heilbronn fell, on 12 April, with a bridge gained over the Neckar. Part of 653 did not withdraw from Schwäbisch hall until 13 April and Stuttgart fell on 20 April.

34.5 Report dated 11 April 1945

Radio-signal: To OKH/Gen. Insp. d. Pz. Truppen
From s.Pz.Jg.Abt. 653,

Ordering: *H. Gr. G rom la Nr. 1592/45 secret Kdos from 7.4.45*

The panzer log s.Pz.Jg.Abt.653, position on 9 April 1945: Total 17 Jagdtigers, 10 operational, 7 needing long-term repairs (over 5 days).

Operational stocks of the battalion by present practice, material situation; carry on requesting. Again requesting urgent delivery of special assignment of 50 cbm spares and immediate conveyance of spares.

Ob.Kdo. H.Gr d.Genst.
i.A. gez. Wilutzky, Oberst i.G.
rom la/. 312/45 Geh.Kdos.
F.d.R.d.A.

Plate 303. Fw Schlabs and Fw Gartner, Schweindorf April 1945 (Karlheinz Münch).

34.6 Withdrawal east for new operations in Nürnberg

After the relative stabilization of the line along the Kocher and Neckar rivers on 10 April, 653 were ordered to move eastward to join defensive force being assembled in the city of Nürnberg. A heavy defense was planned to prevent capture of the city, symbol of the genesis of National Socialism. OKH ordered 653 to move as quickly as possible to this new area of operations. At this time the Allied attacks were making huge advances towards Czechoslovakia.

Some cannibalization had occurred from the seven Jagdtigers needing major repair to get the ten operational, taking five days of intense work by the works company.

The Works Company again commandeered the transport train for the damaged Jagdtigers, with loading taking place at Leonberg and as usual they used heavy camouflage on the rail flatcars. There was now very little other rail activity in the area. For four days the train traveled through Stüttgart, Augsburg, and München to Traunstein near Salzburg.

34.7 Report dated 14 April 1945

Radio-signal: To OKH/Insp. d. Pz. Tr.
From s.Pz.Jg.Abt. 653,

Panzer log s.Pz.Jg.Abt. 653 position on 14 April 1944 1100 hrs

Total 17 Jagdtigers, 5 operational, 6 need short-term repairs, and 6 need long-term repairs

New breakdowns since 10 April - 5

Bergepanther log, total 4, 1 operational, 2 need short-term repairs and 1 needs long-term repairs.

Unit dangerously short of spares, particularly track spares.
There is an immediate pressing new request for:
3 sets of tracks for Pz V
3 sets of tracks for Jagdtiger II B Henschel suspension
2 sets of tracks for Jagdtiger II B Porsche suspension.[2]

34.8 The battle for Nürnberg

The 653 continued to move east as a scattered unit of small groups. One group without Jagdtigers reached Nürnberg on 14 April. The city was the birthplace of National Socialism and had to be defended at all costs.

It is 160 km between Crailsheim and Nürnberg. The Jagdtigers needing long-term repair were being moved by rail.

Five Jagdtigers drove towards Nürnberg but did not complete the journey. In Nürnberg a defensive force of 2 German Divisions, 1 Luftwaffe and a Volksturm Battalion awaited the Americans. In total, there were 35 tanks and Jagdpanzers available to the defenders. The city, ruined by heavy bombing, was ringed by a large number of anti-aircraft guns.

On 16 April, the Allies attacked Nürnberg with two U.S. Divisions. One attacked along the E5, the other along the E6 from Bayreuth. A third U.S. Division from the 21st Corps advanced into the western suburbs through Furth. The attack started under a storm of fighter/bombers hitting the next area of advancement. The defenders fought well and a vast amount of American armor was lost to the German tanks, anti-aircraft guns and Luftwaffe sorties. A U.S. Cavalry group swept around the city to block the escape route to the south.

After three days of vicious fighting, with buildings having to be cleared one at a time by the Americans, they finally broke through the Magineau Line and into the old town on 19 April. The following day, Hitler's birthday, the city finally fell to superior Allied forces.

34.9 Withdrawal south

On 19 April, with the close proximity of the Americans in the southern suburbs of Nürnberg, there was no alternative but to withdraw the 653 elements south of the danger zone. They passing through Steinhausen, on 24 April, and Walchshoven, on 26 April. That day, a strength report was sent to O.K.H. showing 14 Jagdtigers left but with only 1 operational (see later). By now the American advance had become a steady drive through the country with the armor in front followed by support troops.

Because of the difficulty of obtaining spares and the huge amount of work involved getting the Jagdtigers operational, Rolf Fromme decided to try to withdraw the vehicles back to the factory at Linz. He sent a detachment forward commanded by Lt. Hans Knippenberg to make the necessary arrangements and take on the new vehicles nearing completion.

On 27 April, the unit passed through Kissing by Augsburg one day ahead of the Americans. There was no fight for Augsburg, which fell, on 28 April.

The withdrawal now swung east, passing through Wessling, on 28 April and then München/Geiselgasteig, on 29 April, with further movement through Sauerlach to Bad Aibling, on 30 April, the same day that München fell with virtually no fighting.

Throughout all the withdrawal, the train with the six long-term broken Jagdtigers had to be run at night with reconnaissance vehicles moving ahead to plan the following day's rail route. There was always the threat of fighter/bomber attack and damaged rail lines.

34.10 Report dated 26 April 1945

Radio-signal: To OKH/Insp. d. Pz. Tr.
From s.Pz.Jg.Abt. 653,

14 Jagdtigers, 1 operational, 7 need short-term repair, 6 long-term repairs.

34.11 Jagdtiger losses in April 1945

1 April 1945 *24 Jagdtigers (3 operational/22 need repair).*

3 April 1945 *23 Jagdtigers (1 operational/ 11 need long-term/11 short-term repairs)*
Jagdtiger 123 is blown up in Eppingen.
Jagdtiger 114 is blown up in Eppingen.

5 April 1945 Jagdtiger 1?? Blown up at Nordheim.
Jagdtiger 1?? Burns out at Besigheim.
Jagdtiger knocked out near Nordheim

6 April 1945 Jagdtiger 214 is blown up at Lauffen.
Jagdtiger 213 is blown up at Besigheim.

9 April 1945 *17 Jagdtigers (10 operational/ 7 need long-term repairs).*

14 April 1945 *17 Jagdtigers (5 operational/ 6 need short-term repairs/ 6 need long-term repairs).*

?? April 1945 Three Jagdtigers are blown up during the withdrawal southeastwards from the Schwäbisch Hall-Crailsheim area?

26 April 1945 *14 Jagdtigers (1 operational, 7 need short-term repair, 6 long-term repairs).*

[1]At this stage the U.S. intelligence had examined 9 Jagdtigers-5 Henschel and 4 Porsche. Of the Porsche vehicles, 3 were in the Haguenau area.
[2]This confirms that there were still two Porsche Jagdtigers with the unit at this time!

35

Schwere Panzerjäger Abteilung 653 May 1945

35.1 May 1945 synopsis

1 May 653 begin a long withdrawal into Austria. Unit moves from Rosenheim along the E11 road to Schlossberg. A Jagdtiger from 2/653 falls through a bridge in Kolbermoor near Rosenheim.

2 May The Battalion withdraws 30km to Erlstätt (Chiemsee). A detachment under Lt. Hans Knippenberg is on standby to receive orders to take over a new Jagdtiger at Nibelungen Werk.

3 May The 653 Travel 45km through to the battalion command post at Henndorf. Ten of last 13 Jagdtigers were blown up near Erlstätt, the troops move to Henndorf.

4 May No orders had been received to pick up the Jagdtigers at Nibelungen Werk. No gun sights were available for them. Orders were given to blow up 8 Jagdtiger at the factory. This was carried out. The 653 troops travel 45km to Bad Ischl. The Americans occupy Linz and Salzburg. Amstetten is occupied by the Soviets. Jagdtiger No 113 is self-destructed at Rosenheim.

5 May Bulk of 653 in Bad Ischl. The new Jagdtigers drive into Soviet Forces at Amstetten. One Jagdtiger breaks down the other three are withdrawn to Strengberg and surrendered to U.S. Forces.

6 May Part of 653 at Unzenau.

7 May Unit remnants withdraw to Liezen south of Linz where the last two of the original Jagdtigers No 312 and 324 are surrendered to U.S. Forces at 15.00 hours, along with Army Group South.

8 May The unconditional surrender of the German Armed Forces was signed.

9 May Soviet Forces took over the Nibelungen Werk.

10 May Allied intelligence reports are circulated on Jagdtiger.

11 May Dr. Judtmann, Director of the Nibelungen Werk shoots himself.

35.2 The end of schwere Panzer Jäger Abteilung 653

By the end of April 1945, frequent engagement against the Allies together with a long withdrawal and a critical lack of spares had had a devastating effect on the remaining Jagdtigers.

It was mainly the tracks and transmission that were giving problems. The request for spares had not been fulfilled and there had been no time to achieve any significant repair work due to constant movement keeping one day in front of the relentless Allied advance.

Two Jagdtiger, No 113, and a 2/653 vehicle was operational. A Train was still being used for the withdrawal and the objective was now to get the Jagdtigers back to the factory at Linz for repair. Gasoline was siphoned out of the Jagdtigers to keep the lorries and half-tracks available to transport the troops. Advance

Plate 304. Jagdtiger tactical number 113, chassis No 305023, blown up near Rosenheim on 4 May 1945. There were very few spare track links available at this time (U.S. Army).

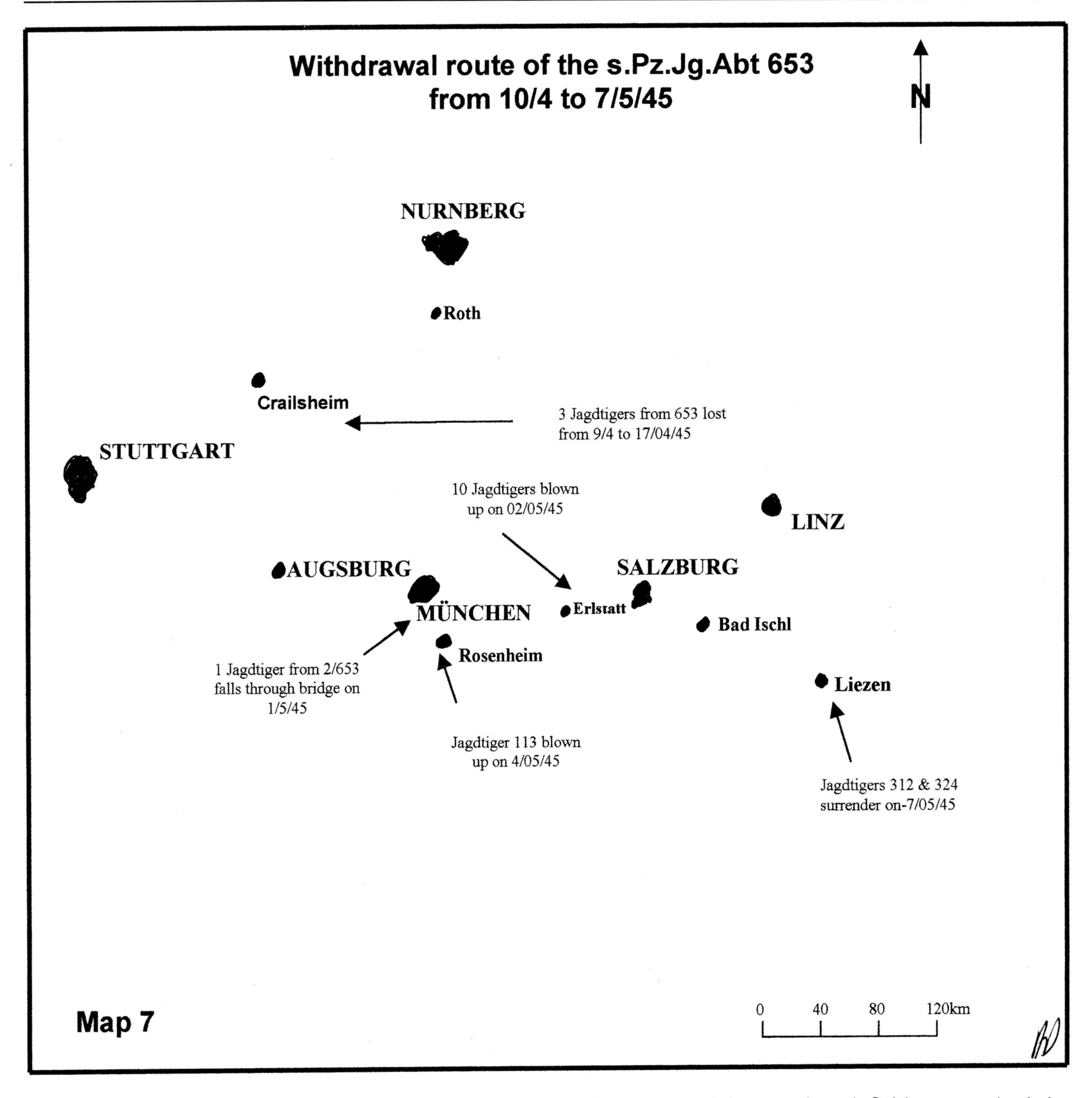

reconnaissance was vital to check the next section of track and rail movement was always in darkness.

On 1 May, part of the Battalion was near Rosenheim with 2 Jagdtigers on road march, the lead Jagdtiger from 2/653 fell through a canal bridge and was left in the water, the second vehicle No 113 chose an alternate route. In the evening the rest of 653 withdrew 30km to Erlstätt (Chiemsee). Jagdtiger No 113 remained in Rosenheim with defective steering until its crew blew it up on 4 May.

On 2 May, Fromme's assessment was:

The pursuing American armor was already threatening Rosenheim about two hour's drive from this position.

The withdrawal due east through Salzburg was also being threatened by a further American advance from the direction of Burghausen.

The railway line to the east was too badly damaged for a quick repair.

Only 2 Jagdtigers were operational.

For these reasons there was no alternative other than to quickly salvage the machine guns, ammunition and the remaining gasoline from the broken down Jagdtigers.

Plate 305. The long withdrawal continues through Austria, in early May 1945 (Karlheinz Münch).

Plate 306. The troops of s.Pz.Jg.Abt 653 drive through Austria (Karlheinz Münch).

Fromme then gave the order to blow up the last ten Jagdtigers, which was skillfully carried out and the Works Company was virtually disbanded before the remains of the Battalion drove away due east to the battalion command post at Henndorf.

Part of the last battalion order dated 2 May 1945 stated:

The battalion is immediately assigned to Army Group South, it will move by road march through Salzburg to Linz, initially holding the Henndorf-Seekirchen and Neumarkt area northeast of Salzburg.

Paragraphs stated:

4) Jagdtigers 312 and 324, including Uffz Reimann's recovery group, 3 maintenance groups and one KFZ 100 with side drives will move to the new area through the night. Group commander is Oblt Kretschmer.

6) The crews assigned to take over Jagdtigers at Linz will remain ready.

Major Fromme

On 3 May, Henndorf was reached after driving 45km along the E11 and through Salzbourg which was threatened by an American attack from the north.

On 4 May, the Battalion withdrew to Wessenbach (Attersee). The Allies occupied Linz and Salzburg on this day, there was no longer any point in proceeding towards Linz.

The withdrawal, on 5 May, was through the beautiful Austrian valleys in perfect spring weather to Bad Ischl. Small American columns were still following.

Plate 307. Sd Kfz 251/8 in Liezen just before war's end, 7 May 1945 (Karlheinz Münch).

Plate 309. The last Jagdtiger crews in Austria, in May 1945 (Karlheinz Münch).

The next day, May 6, they withdrew southeast to Putsches where some of the troops were captured because of broken down transport vehicles. In spite of the steep roads the two Jagdtigers were still mobile.

The final move in this lengthy journey was made, on the night 6/7 May, when the remaining troops and the two Jagdtigers drove to Liezen on the Enns River. At 1500 hours they were surrendered by Rolf Fromme, at the same time as General Rendulic surrendered what little remained of Army Group South to American forces.

Thus ended the operation of the largest Jagdtiger unit, s.Pz.Jg.Abt. 653, one day before the unconditional surrender of the German armed forces in Europe, signed by General Jodl, on May 8 1945.

35.3 Conclusion

Schwere Panzer Jäger Abteilung 653 was a fully self-contained until the end of the war and not under direct Army Group control. Regular reports were being sent to O.K.H.

Comments made by s.Pz.Jg.Abt 653 state that only about 30% of their Jagdtigers were lost to enemy action, the rest were either blown-up or abandoned due to mechanical failures.

Only four of the forty-two Jagdtigers were retrieved intact, all were from 3/653 and one still survives in the USA.

The Battalion in spite of suffering from severe mechanical problems, destroyed well over 200 Allied tanks and vehicles with their Jagdtigers.

Four crews who took on the new Jagdtigers at Nibelungen Werk, in April 1945, were no longer under the direct command of s.Pz.Jg.Abt 653 but under the command of a LAH detachment as part of the 6th SS Panzer Armee (see Chapter 40).

Plate 308. War's end for the troops of s.Pz.Jg.Abt 653 (Karlheinz Münch).

36

Other Units Intended to Receive Jagdtigers

There were three other panzer units who were planned to receive Jagdtigers, they were:

36.1 Panzer Lehr Division

The first unit planned to be equipped with Jagdtigers was the third company of the Panzer Jäger Lehr Division, this was to take place in March 1944. Fourteen Jagdtigers were to be used four per platoon and two for the company commander. This did not materialize because of slower than expected Jagdtiger production. The company was later equipped with nine Jagdpanzer IV Vomag in June 1944.

36.2 Schwere Panzer Jäger Kompanie 614

This was a newly named tank destroyer company, which took over the remaining Elefant tank destroyers and personnel from 2/s.Pz.Jg.Abt. 653 in October 1944. This transfer occurred in Poland under the command of Oblt Solomon. This subsequently caused the formation of a new 2/s.Pz.Jg.Abt 653 company, which was equipped with 14 Jagdtigers in January 1945.

From records, the panzer logs of the 2/s.Pz.Jg.Abt.653/S.H.Pz.Jg.Kp. 614 (Elefant) were as follows:

	Total	*Operational*
20.9.1944	14	14
25.9.1944	14	14
5.10.1944	14	14
15.10.1944	14	14
25.10.1944	14	14
31.10.1944	14	14
25.11.1944	14	14
5.12.1944	14	13
10.12.1944	14	13

On 30 December 1944, the s.Pz.Jg.Kp.614 had fourteen operational Elefants and was at full strength.

They were in action for the first time against the Soviets on 13 January 1945.

By 22 February 1945, the s.Pz.Jg.Kp.614 commanded by Hauptmann Ritter were now down to ten Elephants, they were located in the Sorau area. Order K.st.N.1148 transferred the six Porsche Jagdtigers from s.Pz.Jg.Abt. 653 to s.Pz.Jg.Kp.614, to bring it back too effectively greater than full strength.

On 25 February 1945, s.Pz.Jg.Kp.614 had four Elephants being overhauled, all requiring spare parts. Coincidentally, the six Porsche Jagdtigers of s.Pz.Jg.Abt. 653 were also being overhauled at Bellheim near Landau and transfer was not possible.

3 March 1945, s.Pz.Jg.Kp. 614 were promised reserve reinforcements of 74 officers and men and the 6 Jagdtigers from s.Pz.Jg.Abt.653, the spares for the four Elephants were in transit from Linz.

The objective was:
To bring s.Pz.Jg.Kp.614 back to full strength.

Issue s.Pz.Jg.Abt.653 with six new Henschel Jagdtigers to improve the reliability of the Battalion and standardize on spares.

However the changes did not happen and when, on 22 April 1945, s.Pz.Jg.Kp.614 were in action near Zossen 25km south of Berlin, at the headquarters of the Insp. Gen. D. Pz. Tr., they had only four remaining Elefants. Soviet forces destroyed the company along with the rest of IX Armee.

36.3 SS Panzer Abteilung 501

There was a plan, on 3 November 1944, to equip the third company of SS Pz.Abt 501 with fourteen Jagdtigers. Hitler had wanted a Jagdtiger company with a Tiger Battalion. The Tiger unit was then to be employed in the Ardennes offensive.

Due to production delays, this could not be achieved and the order was changed in favor of creating a new unit, the s.Pz.Jg.Abt.(Jagdtiger) 512, in February 1945.

There is an unconfirmed report that a detachment of 40 men from this unit were sent to Nibelungen Werk in early May 1945, to take on 6 Jagdtigers, this is covered in Chapter 40/7.

37

Schwere Panzerjäger Abteilung 512 January - February 1945

37.1 Origin of schwere Panzer Jäger Abteilung 512

This was a new Panzer Jäger Unit, which started its formation in late January 1945. It was planned to receive 33 Jagdtigers, formed into three fighting companies, each with 10 Jagdtigers and 3 for the battalion commander [1]. This was the second Jagdtiger unit. Previously there has been some confusion as to its correct name, with variations from Pz.Jg.Abt. 512 to s.H.Pz.Jg.Abt.(Jagdtiger) 512, however the most common term used in official records is s.Pz.Jg.Abt. 512 therefore this is used in the text. The terminology used in the individual records is presented as recorded!

At the beginning of January, Major Walter Scherf was appointed Battalion Commander. He was formerly Commander of 3/s.Pz.Abt.503, he then became Battalion Commander of the whole 503 Battalion. He had received the Knight's cross, on 23 February 1944, while fighting in the Tiger I tanks on the Eastern Front. Scherf appointed the first two of his planned three fighting company commanders, in late January. They were:

Hptm Albert Ernst, an experienced Panzer Jäger, he had been awarded the Knight's cross on 7 February 1944, when serving in Nashorns with s.Pz.Jg.Abt 519, he later fought in Jagdpanthers, Ernst commanded the first company 512.

Oblt Otto Carius, from 2/s.Pz.Abt 502, who had recently recovered from almost fatal wounds. Carius won the Knight's cross, on 4 May 1944, and then the Oak leaves, on 27 July 1944, in action with the Tigers of 502 Battalion. Carius commanded the second company.

The Jagdtigers were handed over to 512, at Nibelungen Werk. They were sent for test firing on the ranges, to Döllersheim near Zwettl, a distance of 150km by rail. Throughout late January, 512 personnel were training alongside 2/653, in Döllersheim, before 2/653 was sent out to the Landau-Bellheim area, at the beginning of February.

In January and February 1945, the concentration area for 512 was in Westenholz by Delbruck. The Battalion HQ was situated in the milk dairy of Epping Westenholz, 15km west of Paderborn.

Early in January, Walter Scherf as Commander of 512 was ordered to travel to Nibelungen Werk, to see that all was in order when the Jagdtigers came off the production line, then later into service after their test firing at Döllersheim.

By the end of January, only a small number of the men had been deployed to 512. No Jagdtigers were ready to be dispatched to Döllersheim. However, 11 Jagdtigers were completed at Nibelungen Werk, by the month end. These were undergoing quality inspections by the factory's test engineers.

37.2 Schwere Panzer Jäger Abteilung 512 - February 1945

The equipping and training of 512 continued throughout the whole month of February. Personnel from the s.Pz.Abt.424 (s.Pz.Abt.501) and s.Pz.Abt.511 were seconded for duty to staff the three fighting companies, each planned to have 10 Jagdtigers - 1 for the Company Commander and 3 for each of 3 Platoons.

On 6 February, 512 were ordered by the Fuhrer's HQ, via the Commander in Chief OKH, to be in position with 31 Panzers in the area of Paderborn.

The organization of s.Pz.Jg.Abt 512 was as follows:

Commander Hauptmann Walter Scherf (Ritterkreuz, Deutshes Kreuz in Gold)

Adjutant Lieutenant Hans Senftleben
Ordonnanzoffizier Lieutenant Rolf Kettener
Abteilungsartz Oberartz Dr. Konigshausen
Paymaster Oberzahmeister Otto Brenner

1 Company Oberleutnant Albert Ernst (Ritterkreuz)

1 Platoon Commander Lieutenant Heinz Rondorf
2 Platoon Commander Lieutenant Kubelka
3 Platoon Commander Lieutenant Sepp Tarlach

2 Company Oberleutnant Otto Carius (Oakleaves on Ritterkreuz)

1 Platoon Commander Lieutenant Hans Jurgen Hass
2 Platoon Commander Lieutenant Fritz Riess
3 Platoon Commander Lieutenant Von Salisch

Versorgungskompanie (Maint Supply)	Oberleutnant Hans Fink
Werkstattkompanie (Workshop)	Oberleutnant Dipl. Ing Wetterling

A reference in OKW Volume 8, page 1385, dated 10 February 1945, stated that the Schwere Panzer Jäger Abteilung 653 (32 Jagdtigers), also the schwere Panzer Jäger Abteilung 512, 2 Company [2] expected to be ready for action by 22 February in Döllersheim.

37.3 Order 11.2.45

High command of the army Gen.Insp.d.Pz.Tr. Berlin 11.2.45 ObdE/AHA/Staff Ia (1)

Nr.9708/45 secret.
Ordering: OKH/Gen StdH/Org.Abt. Gen.Insp.d.Pz.Tr. /Org II Nr 570/45 gKdos.
and II Reports
Operation: List s.Pz.Jg.Abt (Jagdtiger) 512.

Through W.Kdo. VI, it is with support from the Arb. Stb. (Bz) to Tr. Ob.Pl. Sennelager the s.Pz.Jg.Abt (Jagdtiger) 512. Short duration to send. Arranged dispatch day for:
1.Kp. with accompanying supplies and repair-service - 15.2.45.
Staff, Staff Kp. and 2.KP. along with support and repair service - 25.2.45.
Rest of the battalion with 3.Kp. - 5.3.45.

Organization and strength.
Staff and Staff Kp. Secret. KStN (fG) from 1.11.44, Ausf.D, with changes:
to drop: 1Kdt. (St. Gr. G)
2 Direct guards (St. Gr. G)
2 Pz drivers (St. Gr. G)
2 Track-guards (St. Gr. M)
2 Jagdtigers[3]

to add: I Radio operator (St. Gr. G)
3 Radio operator (St. Gr. M)
2 Zgkw. drivers (St. Gr. G)
2 m. Beob. Pz.Wg. (Sd.Kfz.251/18)

3 x s.Pz.Jg.Kp. (Jagdtiger) Secret. KStN 1176 (fG) from 1.11.44.
(each Kp. 10 Jagdtigers) each Kp. 12 track-guards (St. Gr. M) additional.
Support. Kp. s.Pz.Jg.Abt Secret. KStN 1151 b (fG) from 1.11.44.
To further explain:
Page b, line 33 - 1 s.Lkw.
Page c, line 36 - 5 s.Lkw.
Page d, line 7 - 5 s.Lkw.
line 19 - 1 m.Lkw.
line 34 - 1. m.Lkw.

Pz. Works. Kp. Secret. KStN 1187 b from 1.11.44.
Until the issue of new KAN apply for:

Staff, Staff Kp.	*KAN*	*1107b (fG) from 1.6.44.*
s.Pz.Jg.Kp. (Jagdtiger)	*"*	*1176 (fG) from 1.6.44.*
Support. Kp. (Jagdtiger)	*"*	*1151b (fG) from 1.6.44.*
Pz. Works. Kp	*"*	*1187b from 1.7.44.*

For the list, standby for order:

All 1.Kp. personnel unit "Buschmann" (s.Pz.Kp.Tiger) with accompanying support units and repair - service.

The units current personnel composition, are through W.Kdo. XVII, each without fighting troops and repair - service, are immediately to move from Tr.Ub.Pl. Döllersheim to Tr.Ub.Pl. Sennelager.

The fighting troops and repair - service, through AHA/In 6, for taking over the battle-issued Jagdtigers and proceed with Jagdtigers to Tr.Ub.Pl. Sennelager.

s.Pz.Abt. 424 (without personnel for 1.Kp. they now belong to support unit and repair - service).
s.Pz.Abt. 424 (without 1.Kp) with this, it disbands.

Personnel regulations:

Deficiencies in personnel to be made up from Pz.Ers.u.Abt. 500 Paderborn.
Liberated personnel (including 1. /s.Pz.Abt.424) are to transfer to Pz.Ers.u.Abt. 500 Paderborn.

Training Period through Insp.d.Pz.Tr.

All replacement troop sections must come from Pz. Ers.u.Ausb. 500(W.K.VI)

Other appointment performance regulations. Order AHA/Staff Ia Nr. 1/45 Gkdos. From 1.1.45.

The first 5 Jagdtigers released to s.Pz.Jg.Abt 512 were recorded as released on 30.1.45 through H.Z.A. Linz and taken on to Döllersheim on 12.2.45, a further 6 were released from the factory on 14.2.45 to H.Z.A. Linz, after subsequent recall of these 11 vehicles, the release entries were lined out.

On 16 February, an order was sent by OKH delaying the equipping of 512 because of serious mechanical problems with the steering units. The same day, the order was issued to change all steering units after design changes had been completed, 5 Jagdtigers were already going through firing practice, in Döllersheim, and would now need further firing trials after future repair in Nibelungen Werk. (Cannons would have to be removed to install the new steering units).

This new Jagdtiger unit, which was on schedule in compliance with ObdE/AHA/Staff Ia(1)Nr.9708/45 secret would now be considerably delayed!

37.4 Interruption to the Jagdtiger production

A serious weakness was found in a design fault with the Henschel L801 steering unit. A report prepared for the General

Inspector of the Panzer Troop General Major Thomale by Oberst Crohn of the Motorisierung/Preuf. Pz Kummersdorf:

Damage on steering unit gearing on the Jagdtigers. Report by Oberst Crohn, testing the tanks on the 16 February 1945. There are mass production faults on the steering units of Jagdtigers, which were showing after road distances of 250-400km. The Inspector General's decision was:

1. Jagdtigers not to be sent to the troops until the new steering units had been installed.

2. The Jagdtigers already in action were to be pulled out, one at a time, for the alterations on the steering. This was to be done by the Workshop Company after completion of all documentation and preparatory work.

Gen. Insp.d.Pz.Tr. /Lt.Kf.Offz.
Bd.Nr.3570/45
geh.v.16.2.1945

De Chef des Stabes
Generalmajor Thomale

Concern: employment readiness s.H.Pz.Jg.Abt. (Jagdtiger) 512

1. Operation of the Jagdtiger from the s.H.Pz.Jg.Abt.653 acquired serious damage on the steering gear housing (construction weakness).

2. It is necessary to change the unit on all Jagdtigers.

3. Design changes to be made first on all Jagdtigers that were still in production in the factory.

a. 5 Jagdtigers of the s.Pz.Jg.Abt.512 in Döllersheim
b. 6 Jagdtigers in the factory at Linz

4. The deployment of s.Pz.Jg.Abt.512 will be delayed and you will be told the time and date of the design change when it is known.

Gen. Insp.d.Pz.Tr./Lt.Kf.Offz.
Abt. Org.IINr.746/45 v 16. 2. 45

Der Chef des Stabes
Generalmajor Thomale

This serious mechanical fault severely reduced the operational reliability of s.Pz.Jg.Abt.653 and incurred major delays to the equipping of s.Pz.Jg.Abt.512.

37.5 Contingencies due to delays

On the 18 February, Walter Scherf made his last visit to Döllersheim to see 2/512 (Carius) on the ranges where they were sighting in the guns. Scherf took the opportunity himself to shoot off 3 rounds with the 12.8cm cannon, and was extremely impressed by its performance. He recognized it would be some time before his Jagdtigers would now become available.

On the evening of 21 February, Scherf returned to Westenholz with Fw Schröder his driver. Walter Scherf sent Obfw Heinz Grien, an experienced tank driver and Tiger Commander to Nibelungen Werk to continue to oversee the production and testing of Jagdtigers through March and April. He reported directly to Walter Scherf.

Official German tank (production figures) indicates that Nibelungen Werk built 13 Jagdtigers in February. These were assigned to 512 before completion, hence the 6 February figure of 31 Jagdtigers to 512.

Throughout February, Otto Carius was assembling his personnel at Sennelager and Döllersheim where the armored unit was being formed. He was constantly commuting back and forth the 1000km between the 2 locations until his Jagdtigers finally arrived at Sennelager in March. Full training was then able to commence for the whole of 2/512. However, events elsewhere would quickly disrupt this.

The first Company to be ready was actually second Company (Carius), the next was first Company (Ernst) and the last third Company (Schrader)

To save transport time, the combat tracks for the Jagdtigers were delivered straight from the factory to Sennelager.

37.6 Jagdtiger deliveries in February 1945

Delivered to	Transported on	No	Type	Chassis No	Tactical No
Döllersheim	*12/02/45*	*5*	*H*	305050	?
			H	305051	?
			H	305052	?
			H	305053	?
			H	305054	?
s.Pz.Jg.Abt.512	*14/02/45*	*6*	*H*	305055	?
			H	305056	?
			H	305057	X5
			H	305058	X7
			H	305059	?
			H	305060	?

Note: - Order on 16/02/45, return these 11 Jagdtigers to factory for new steering units.

37.7 Jagdtigers built by Nibelungen Werk in February 1945

There were 13 Jagdtigers completed in February against a program of 20.

Chassis No:305062
305063
305064
305065
305066
305067
305068
305069
305070
305071
305072
305073
305074

37.8 Hulls built by Eisen Werk in February 1945

There were *7 Hulls completed giving a running total of 112.*

Bomb Attack on factory 17/Feb/45, - 12 Tons HE.

(1)Secret. KStN 1176 (fG) from 1.11.44. This would be revised on 6/2/45, and again on 11/2/45 see following documents.
(2)This Company did not actually complete its preparation until the second week in March. Its 10 Jagdtigers finally arrived in Sennelager, on 8th March.
(3)This dropped the planned total number of Jagdtigers for the unit down to 30.

38

Schwere Panzerjäger Abteilung 512 March 1945

38.1 Combat diary March 1945

1 March No fighting companies are ready, Jagdtigers delayed due to steering units.

3 March The first 5 Jagdtigers of 2/512 leave Linz by rail for Sennelager.

5 March 5 more Jagdtigers leave Linz for Sennelager. The first 5 arrive at Sennelager.

7 March The second 5 Jagdtigers arrive at Sennelager on the night of 7/8. U.S. capture bridge at Remagen, in the afternoon, word reaches Sennelager.

8 March 2/512 start full training, the Jagdtigers start firing trials.

10 March 1/512 Jagdtigers sent from Linz to Sennelager. 2/512 having problems with guns/optics. Hitler orders 512, to attack at Remagen, they are not ready.

13 March The 10 Jagdtigers of 1/512 arrive in Sennelager that night. Ernst's company starts full training.

14 March 2/512 with their 10 Jagdtigers are loaded onto 3 trains destined for Siegberg: They have had 1 week's full training! 1/512 Jagdtigers go to firing ranges.

17 March Remagen Bridge falls into the Rhine, the Americans have other bridges.

18 March Carius arrives in Siegberg, 5 of his Jagdtigers in Duisburg!

19 March 2/512 arrive at Siegberg, an attack is not possible. 1/512 load at Sennelager.

21 March 5 Jagdtigers of 3/512 are sent by rail from Linz to Sennelager.

23 March 2 Jagdtigers of 3/512 are sent from Linz to Sennelager. Nibelungen Werk is heavily bombed. British Army crosses the Rhine near Wesel.

25 March 5 Jagdtigers of 3/512 arrive at Sennelager. U.S. attack north west of Remagen.

26 March The 3/512 commence training at Sennelager, that night 1/512 unload at Olpe.

27 March 2/512 travel to Siegen. 2 Jagdtigers were lost through stupidity!

28 March 1/512 attack from Siegen towards Dillenburg.

30 March 3/512 with 5 Jagdtigers are on Sennelager ranges. 2 more of 3/512 Jagdtigers cannot reach Sennelager.

31 March 3/512 ordered into action against U.S. Forces converging on Lippstadt. 1/512 are ordered to be part of the breakout attempt near Oberkirchen.

38.2 Schwere Panzer Jäger Abteilung 512's Jagdtiger repairs

When the order to change the steering units was issued to Battalion, on 16 February 1945, the Jagdtiger deployment to 512 was as follows:

Döllersheim: ***5 Jagdtigers*** (chassis no 305050 - 305054)

Linz: (Army reserve) ***6 Jagdtigers*** (305055 - 305060)

Order to return these eleven vehicles to Nibelungen Werk, to change the L801 steering units. (After their return, the entry on the issue record was lined out).

On 1 March 1945, the Panzer log of the Ersatzheer, records, six Jagdtigers still not transported back for repair.

On 6 March 1945, three Jagdtigers were still to transport back to Nibelungen Werk for modification. The transport requested since the 24 February 1945.

At the factory, with the 150t traveling overhead cranes it was a relatively easy operation to remove the main armament. Production line workers removed the old steering units. Segregation prevented further accidental installation, followed by a return to Henschel, for refurbishment. Nibelungen Werk notified Insp. Gen. D. Pz. Tr., they would need approximately 5 days to change the units on the 11 Jagdtigers, already released to 512, on their return. The main difficulty would be the rail transport, the air raid on 17 February, had damaged the shunting-yard and also many of the special flatcars.

On arrival of the new steering units, chassis No's (305062-305070) were the first modified. These were still on the production line that had been stopped and prepared for the modifications. After the notification from the Insp. Gen. D. Pz. Tr., in mid-February, Dr. Judtmann had been in regular contact with Waffen Pruef 6, and Dr. Ing. Stieler Von Heydekampf, his counterpart at Henschel.

A separate area adjacent to the production line was prepared for the work on the Jagdtigers returned. The defective steering units were a significant delaying factor to Jagdtiger production, in February 1945.

At the beginning of March 1945, none of the three fighting companies were ready for action. On 5 March, OKH received the following report.

38.3 Status report, s.Pz.Jg.Abt.512 on 5 March 1945

Notice for Fuhrer's report on 5.3.45.
Position of the inventory Jagdtiger - Abt. 512. Sennelager:

Personnel: Full
Jagdtiger provision:

2/Kp: 5 Jagdtigers dispatched at 1200hr from Linz on 3 March to Sennelager. 5 Jagdtigers to send at 1200hr on 5 March from Linz to Sennelager. 10-rail transport, the transportation expected to take about 3 days.
10 Jagdtigers
1/Kp: 3 Jagdtigers finished in factory. 5 Jagdtigers in factory without tracks, 2 further Jagdtigers will be completed in factory by the 8 March.
10 Jagdtigers

3/Kp: 1 Jagdtiger in factory will be completed by 8 March, 6 Jagdtigers will be completed by 10 March.
7 Jagdtigers
The combat tracks for all the Jagdtigers should have been delivered direct to Sennelager from the August Engels factory.

38.4 Status report Jagdtiger - Abt 512 on 6 March 1945

2.Kp.: On 3.3.45 5 Jagdtigers on transport from St. Valentin
On 5.3.45 5 Jagdtigers on transport from St. Valentin
10 Jagdtigers

1.Kp.: 5 Jagdtiger in factory finished. Field optics. Courier expected from Jena on 7.3.45.
3 Jagdtiger in H.Z.A Linz transport for their return requested since 24.2.45. Transport not possible, the transport coordination not in a position to acquire S syms wagons (flat cars). 35 S syms wagons blocked, in part destroyed by bombing, 2 Jagdtigers in St Valentin will be finished on 8.3.45.
10 Jagdtigers

3.Kp.: 1 Jagdtiger on 8.3 in St Valentin will be finished. 6 Jagdtigers, Z. zt in Linz for steering units need changed. Time requirement, about 5 days, earliest possible finish 10.3.
7 Jagdtigers

Altogether 27 Jagdtigers, all with 12.8 guns.[1]
30 sets of tracks transported to Sennelager. Z. zt in St Valentin, Bzw. Linz wanting transport tracks, these needed immediately.
Ammunition: 3100 rounds available 25% are Pz.GR and 150 rounds for test firing. These already cleared in Godenau Harz in prepared frames. Firing of the test ammunition and other ammunition, directly after Sennelager requisition. No more cannons in stock or mount parts available from firm in Breslau area [2]*. No Jagdtigers have been test-fired so far. Test firing is compulsory in Sennelager, consequences not known.*

38.5 The bridge at Remagen

On 7 March 1945, the Americans near Remagen captured the Ludendorf Bridge; it was only slightly damaged by a failed demolition attempt. Attacking American troops had thrown the demolition charges into the river.

This was the major strategic loss at this stage in the war in Western Europe and affected every German combat unit.

The War Diary record of OKW on 7 March records:

"The Enemy had reached Kreuzberg and as far as a bridge near Remagen, which it appears was encumbered with fugitives. They crossed the bridge and succeeded in forming a bridgehead on the eastern bank of the river, counter-attack early this morning. The 11th Panzer Division will be brought from Bonn but gasoline is in short supply."

Twenty-four hours after the first crossing, 8,000 Americans were in the bridgehead. Repeated German attempts to destroy the bridge with bombs, artillery and even a railway-mounted gun, all failed.

The bridge finally fell into the river, on 17 March; a Bailey Bridge had already replaced it.

38.6 More troops for s.Pz.Jg.Abt.512

On 13 March, s.Pz.Abt.511 (formerly s.Pz.Abt.502) was reorganized. 2/511 took over the Battalion's remaining Tigers, 1/511 were re-equipped with Hetzers. The remaining Panzer crews of 511 were transferred to complete the formation of 3/ s.Pz.Jg.Abt.512, they were also training in Sennelager.

Later between 20-25 March, Oblt. Schrader was appointed company commander 3/512. Further troops from the 3/ s.Pz.Abt.501 (s.Pz.Abt.424) who in early February 1945, had lost their Tigers near Lisow and were transferred to Sennelager, in early March, to join Pz.Ers.u.Abt. 500 and later form 3/512 [3].

38.7 The battle for the Ruhr pocket

With the failed offensive in the Ardennes, Hitler had lost most of the armor he required for the defense of Germany. The next natural defense barrier was the River Rhine, the last defense of the Ruhr (the workshop of Germany) and a great psychological barrier to both sides. Both the Allies and Germany were prepared to fight bitterly for the Rhine. The main assault was to be made in the north by the British and Canadians against the Ruhr, but by a stroke of luck, the Americans were first across the Rhine when they captured the Rail Bridge at Remagen, on the 7 March 1945.

The assault on the Ruhr was started on the night of 23 and 24 March, when the British, Canadians and the U.S. Army Units crossed the Rhine near Wesel.

The stage was set with the two bridgeheads over the Rhine, for a classic pincer movement to entrap German Army Group B (Model), in the Ruhr region.

The pocket was sealed at 1300 hours, on 1 April 1945, when the U.S. First and Ninth Armies linked up at Lippstadt.

The number of soldiers trapped exceeded both Stalingrad and Tunisia, with the ultimate surrender of captured Generals and Divisions looking like a list of Germany's best.

Because of the low morale of the battered Wehrmacht, the number of casualties for such a large operation was surprisingly low. Some units, notably the schwere Panzer Jäger Abteilung 512, without friendly air cover, and left to the mercy of the Allied fighter bombers, were able to scrape together a small force of tanks and self-propelled guns (Jagdtigers) and fought with unbelievable ferocity and zeal at this stage of the war.

The Battle to eliminated the Ruhr pocket took just over two weeks to reach its inevitable conclusion.

38.8 s.Pz.Jg.Abt.512 thrown into action

The first of the fighting companies to near readiness for action was 2/512 (Carius). At the beginning of March, most of his men were assembled at Sennelager, awaiting the arrival of their Jagdtigers from Linz.

The 2/512 were planned to be the first fighting company to become combat ready and they should have reached this state by the middle of February, the need to change the faulty Jagdtiger steering units caused this delay.

Five of the Jagdtigers first assigned to this company had been test-fired on the ranges at Döllersheim where both Carius and his Battalion Commander (Walter Scherf) experienced at first hand the supreme power of the Jagdtiger's main armament.

On 5 March, their first 5 Jagdtigers arrived at Sennelager followed 2 days later, by a further train with their other 5 Jagdtigers. The 12.8cm ammunition had been sent direct from Magdeburg. All the Jagdtigers arrived with their transport tracks fitted, the battle tracks having been sent to Sennelager from the August Engels factory at Velbert.

The 2/512 company was made up as follows:

Company Commander - Oblt Otto Carius	1 Jagdtiger
1 Platoon Commander - Lt. Hans Jurgen Hass	3 Jagdtigers
2 Platoon Commander - Lt. Fritz Riess	3 Jagdtigers
3 Platoon Commander - Lt. Von Salisch	3 Jagdtigers

On the night of 7/8 March, the news reached Sennelager that the Americans had captured a bridge intact at Remagen that afternoon. This panicked the Germans, as the Rhine was to be the last great defense line. This put a panicky haste into 2/512's training because none of 1/512's Jagdtigers had arrived at Sennelager.

During what was to become a very short and intense training period 2/512 experienced their first failure with the Jagdtigers. Most of 2/512's crew members had been seconded for duty from s.Pz.Abt.502 and were used to the relative reliability and flexibility of the Tiger I tank. They were familiar with engaging targets by the easy rotation of tank turret through the full 360 degrees.

The Company Commander, Otto Carius, did not favor this new vehicle. He had great respect for the Tiger 1, with which he had achieved a tremendous success and reputation. The men sensed his dislike of the Jagdtiger and the company was not confident.

The huge Jagdtiger did not perform to the former Panzer crews expectations. They were very satisfied with its armor but not its maneuverability. Any traversing of the gun beyond 10 degrees either side off center necessitated full slewing of the whole vehicle. This soon put the transmission and steering units out of order, as pointed out by some of the 2/653 Battalion's drivers in Döllersheim. The external 'A' frame (travel lock) for the 12.8cm gun had to be released from outside the Jagdtiger, at the very last moment, this put the crews at risk even from small arms fire. The heavy gun barrel, total weight 8 metric tons, needed correctly securing during driving. This was necessary because the aiming gear would have been worn out too quickly. This would have prevented exact aiming and hence this weapon designed to destroy tanks at very long ranges would have been useless.

During training at Sennelager, the Jagdtigers' guns were finally calibrated on the firing ranges. Throughout the first firing trials with the test ammunition, the Jagdtigers missed everything and the crews soon got fed up! Something was obviously wrong and an ordnance technician had to be called in. He quickly solved the problem. The s.Pz.Jg.Abt 512 soon discovered that even after only a short run off the road the gun was so knocked about that its alignment was not the same as that of the optics. Understandably, the men were not confident with their new secret weapon with so many things going wrong before they had seen action.

The problems with the guns and transmission units retarded progress during training and the eventual battle readiness of this company. They knew full well that Hitler considered his Jagdtigers to be one of the secret weapons that could still win the war for Germany. The company commander was not at all happy with the standards of training of his Jagdtiger crews but they had been ordered by Hitler to make an attack on the Remagen Bridgehead, on 10 March. It would later become apparent that Field Marshal Model was not informed of this order.

On 13 March, two trains arrived at Sennelager bringing the 10 Jagdtigers of 1/512. Carius knew that the following day he would have the 3 trains required for conveying his fighting company to the action area. That night he gave his troops an evening pass after they had got all their vehicles to the station and camouflaged them. Only a few of the support vehicles were available to make the first journey with his company.

On 14 March, 2/512 were loaded onto rail transport at Sennelager, their destination was the railway station, in Siegberg.

From there, they were intending to make an attack on the American Bridgehead. At the train station in Sennelager, all 10 of the Jagdtigers and some of the support vehicles were loaded according to plan, the Jagdtigers were put back onto transport tracks for the journey. Fortunately, there was no Allied air attack at this time, even though all the Jagdtigers and support vehicles were assembled at the station.

Because of the constant threat from Allied aircraft and the damage being inflicted on the rail network, the commander continuously drove ahead of the trains in a Kubelwagen. The threat from Allied aircraft was so serious that the trains would only travel at night. For the German troops, it was extremely frustrating how easily and how destructively these aircraft, could run amok with their transport system and weapons's systems. Throughout the days, the trains were either hidden in tunnels or parked against protective slopes. This was the safest they could be kept under the circumstances. During daylight hours, smoke from a steam train could be seen for miles by the Allied air-crews, trains presented easy targets for their rockets and machine guns.

After many diversions, it took four nights for 2/512 to cover the 200 kilometers along the railway from Sennelager to Siegberg, one train was caught in an air attack and the support vehicles were slightly damaged. When Oblt Carius was certain his trains could get to Siegberg station in the early hours of the morning on 18 March, he drove forward to check out the unloading ramps. The Americans were already firing on these positions with artillery. After reporting to the Liaison Officer, Otto Carius discovered that no one within the command structure of Army Group B knew anything about their deployment in this area.

Everything that could go wrong seemed to. In Siegberg, one of the motorcycle riders proudly told Otto Carius that his first train was unloading in Duisburg, 100km north west of where they should have been! The rider was immediately ordered to go to Duisburg as quickly as possible to stop this and correct the situation.

Because of the heavy artillery fire, the first train could not unload in Siegberg that day and stayed in a tunnel until darkness. On the evening of 20 March, the train was unloaded with everything being driven into the woods east of Siegberg and diligently camouflaged, this was vital! All the wheeled supply vehicles had had their tires shot up during an air attack directed at the train, this took several days for the repairs and restricted immediate movement of the whole fighting company. Only 5 Jagdtigers were unloaded, their tracks were changed in the woods, the other train was still in transit from its unplanned diversion!

The Americans had been across the Rhine for 2 weeks and had established a lengthy Bridgehead south of Remagen towards Mainz.

An attack on this Bridgehead by 2/512 alone was now impossible as it was far too well established. They were soon integrated into a larger attack plan.

38.9 The German counter attack

Field Marshal Model was trying to assemble as large a panzer force as possible to make an attack on the north end of the U.S. bridgehead.

On 20/21 March, 2/512 with their first five Jagdtigers unloaded near Siegberg, they were put under the direct command of the 15th Army along with s.Pz.Abt. 506 and s.Pz.Jg.Abt.654. The following day these units were named Panzergruppe Hudel. The assembly area was in the woods of the Lauscheid, south east of Eitorf.

In the early morning of the 24 March, this new Panzergruppe attacked the U.S. bridgehead 9km north of Remagen. The Americans had expected a German attack and had built a strong defense system with anti-tank guns, tanks. This supported with air attacks being directed at any German ground offensive actions, or observed troop movements.

The attack had very little success and was aborted after only a few costly kilometers of progress. The five Jagdtigers were only used as a rear guard, when withdrawal became imminent. The same day the British crossed the Rhine near Wesel, this put more pressure on Model's Army Group B.

On 25 March, Panzergruppe Hudel was forced to withdraw from Siebenbirge, to the west of Wegerbusch, when the American 9th Army launched its attack north west from its positions in the Remagen bridgehead.

Model recognized the threat and expected the key town of Siegen to become one of the Americans first objectives, because from Siegen, many roads run straight into the heart of the Ruhr area. Model ordered General Bayerlein's LIII corps, which included 2/512, to proceed east along the River Sieg and reach Siegen as soon as possible.

On 26 March, both Allied pincers were making excellent progress. The train with 2/512's other five Jagdtiger and two trains with ten Jagdtigers from 1/512 were unloaded near Olpe. The five Jagdtigers of 2/512 in withdrawal were driving along the Sieg River roads. One Jagdpanther and two Tiger II's were blown up before the Panzergruppe crossed the Sieg at Wissen. The 2/512, still with five Jagdtigers in this area, were released from LIII corps and continued to drive towards Siegen.

38.10 Order dated 23/3/45

OKH / Gen St d H / Org. Abt. (Zeppelin)
/ /Org. Abt. (Olga)

Message. : / Ob. d E / AHA / In 6

Gen. Insp.d.Pz.Tr. intended list one Pz. Fla-platoon (without Kfz.) for s.H.Pz.Jg.Abt.512.

Disposition KStN. 1196 and supply to the battalion in Sennelager.

Arrangement: (only P. E.) Pz. Fla - Ers. -u. Ausb. Abt. 1, Wechmar.

Transit arrangements:
Personnel supplied Ob. d E.
Material
4 - Fla-Pz. 3.7 cm single (already arranged)[4]

4 - Fla-Pz. 2 cm four barrel (allocation through Gen. Insp. d. Pz. Tr. each resulting after acquisition from manufacturer).
Weapons and equipment through Ob. d E.
Kfz. Through s. H. Pz.Jg.Abt. 512.

Arranged dispatch day: 1.Halfplatoon: 26.3.45
2 " : After supplying the remaining Fla-Pz.

Plate 310. Side view of blown-up Jagdtiger from 2/512 near River Sieg (Delta Publications).

Plate 311. Further view shows it has been self-destructed after its track broke (U.S. Army).

On the given command

The Gen. Insp. d. Pz. Tr.

Command relay/ Org. II Nr. F 1248/45 g.v.23.3.45

The Chief of staff

(Thomale) Generalleutnant

Teleprinter

Qu., the 29.3.45

Telepr. : Olga 5.47

Message

Reference. : Pz. Fla-Platoon for s. Heer. Pz.Jg.Abt. (Jagdtiger) 512.

1.) Request for short term one

Pz. Fla-platoon for s. Heer. Pz.Jg.Abt. (Jagdtiger) 512

After KStN 1196 sent out and to supply the battalion.

2.) Officer place filling ordered OKH/PA

Material arrangements.

Allocation the gp. vehicle resulted through Gen. Insp.d.Pz.Tr.

Provision with Kfz. through the battalion.

Request for, short term to supply of the remaining material.

OKH/GenStdH/Org.Abt.Nr.I/22878/45 secret. J.A.J.V.

(Litterscheid)

Oberstleutnant i.G. und Gruppenleiter

38.11 Inexperience and lack of training brings disaster

Early on in the combat duration (27 March 1945) two Jagdtigers from 2/512, were in a well-camouflaged position on the outskirts of a wooded area. A long column of Allied tanks at a distance of 1-2km drove straight past them, heading towards Dillenburg. It would have been an ideal chance for the Jagdtigers from 2/512, to have their baptism of fire! Not a single shot was fired! The officer in charge of the two Jagdtigers justified his reason for not firing, because he would have revealed his position. The American tanks would have caused no problems to the Jagdtigers at that range. A heated argument occurred amongst the crews, but no shots were fired!

To make matters worse, shortly after this incident, the same officer ordered his vehicle out of the woods, which really exposed his position. Fortunately, there were no enemy aircraft in the area. The commander pulled his Jagdtiger out without even telling the other Jagdtiger's commander, who promptly followed. Both vehicles raced off like bats out of hell with nothing in pursuit. This careless driving quickly disabled the second vehicle, but the officer did not bother about this, and kept going until his own vehicle broke down. Both vehicles were beyond recovery and were blown up by their crews.

A rumor had circulated Battalion HQ, in the barracks in Siegen, that 2/512 had destroyed 40 American tanks [5]. On arrival Oblt Carius sobered them up, he reported that they had still to achieve their first kill.

The 2/Company with their 3 Jagdtigers, retreated along the River Sieg, through Wissen, Bertzdorf, Kirchen and onto Siegen, which was to be held as long as possible.

The 2 Jagdtigers lost were reported as damaged by fighter-bombers and blown up.

Between 26 and 30 March, 2/512 assembled their remaining 8 Jagdtigers in the Siegen area, the train from Duisburg finally unloaded on the 26 March, 20km from Siegen at Olpe.

The support and works companies were placed in woods outside Geiswald; two Jagdtigers suffered short-term mechanical breakdowns on route and were towed to the works company position for repair.

In Siegen, Walter Scherf himself took command of 2 Jagdtigers from 2/512.

38.12 Fuhrer's Report, 26 March 1945

Ref: s.H.Pz.Jg.Abt.512 Jagdtigers, from General Inspector Tank Troops.

2/Company with 10 Jagdtigers. On 14 March, Jagdtigers were transported from Sennelager. 5 Jagdtigers arrived in the area of Gummersbach and were in action. The Panzer log 25 March showed 3 operational, 2 short-term repairs. The other 5 Jagdtigers [6] were in the area of Olpe ready to be unloaded on 26 March.

The 1/Company with 10 Jagdtigers, together with the Support, Workshop and Supply Companies were ready to be moved from Sennelager on the 19 and 20 March. On 26 March, they arrived in the area of Olpe.

The 3/Company with 5 Jagdtigers arrived in Sennelager on 25 March. 3 other Jagdtigers in St Valentin were ready to be transported on 26 March. 1 Jagdtiger was slightly damaged during the bombing of Nibelungen on 23 March. Repairs on this will be carried out until the 31 March.

38.13 The 1/s.Pz.Jg.Abt.512 go into action

At the beginning of March 1945, 1/512 was with 2/512 in Sennelager awaiting the arrival of their Jagdtigers from Linz. The original plan was that they should have been battle ready by the end of February. The delays with Jagdtigers had put them a month behind schedule.

The report, on 5 March shows three of their Jagdtigers in the factory ready, 5 were ready but without tracks and a further 2 expected to be completed by 8 March. At this stage, none of these Jagdtigers had been on the ranges for test firing.

Albert Ernst and most of his men had been seconded for duty from the Pz.Jg.Abt.519 and were very experienced at fighting with tank destroyers; his most recent fighting experience had been in Jagdpanthers. He had been seriously wounded, before that, he had fought in Nashorns and Jagdpanthers. All his combat experience had been on the Eastern Front.

The 2/512 Jagdtigers had arrived at Sennelager at the end of the first week in March and some of Ernst's men took the opportunity of training with them. Their Jagdtigers arrived on the evening of 13 March, the day before 2/512 were shipped out of Sennelager to make an attack on the Remagen Bridgehead.

Ernst and his crews also had less than a week with their 10 Jagdtigers to train and shoot the guns in. Earlier problems with 2/512's Jagdtiger optics were not too much of a problem, as the lesson had already been learnt.

1/512 were organized as follows:

Commander Oblt Albert Ernst	1 Jagdtiger (No X1)
1 Platoon Lt. Heinz Rondorf	3 Jagdtigers (No X2, X3, X4)
2 Platoon Lt. Kubelka	3 Jagdtigers (No X5, X6, X7)
3 Platoon Lt. Sepp Tarlach	3 Jagdtigers (No X8, X9, X10)

All 3 platoon commanders were very reliable and very experienced officers, who had already served with Albert Ernst, fighting against the relentless Soviet tank assaults.

Unlike 2/512, the troops and commanders of 1/512 were very enthusiastic with the performance of this new vehicle, it offered far more protection than the Jagdpanther.

The 1/512 was loaded onto trains at Sennelager, on the night of 19/20 March. The equipment and troops of the Support Companies and Workshop Companies were also loaded at the same time and occupied the cars between the Jagdtigers to help distribute the weight along the trains. The usual air raid precautions had also to be taken with these trains.

Towards the end of March, the Allied air activity on the German railway system had been intensified to inflict maximum destruction prior to the Allied attack designed to surround the Ruhr industrial complex.

The 1/512 was in rail transit for 6 days. The journey had taken so long because of many diversions due to track damage through air attack, much of which had to be repaired. In addition, because of the changed situation near Siegburg, they were redeployed to the Olpe area. The 1/512 arrived at Olpe, on the night of 26 March, where they were subsequently unloaded.

On 26 March, before his trains had unloaded, Oblt Ernst reported to HQ at Siegen. He was made aware that the attack by the 15 Army had failed against the Remagen Bridgehead. This had been due to the piecemeal application of fighting units as they had been steadily delivered to the area.

In Siegen, 1/512 was given the job of covering the rear of the retreating Harpe Army from its positions north of the Wester Wald. The attached infantry were to be the Freikorps Sauerland.

On 27/28 March, the Jagdtigers drove from Olpe to the south through Siegen towards Dillenburg. When the Jagdtigers carried out their first move to stop the leading U.S. tanks the Freikorps Infantry, which were made up of regional militia units, huddled right up at the back of the Jagdtigers. They were repeatedly warned by the Jagdtiger commanders not to get so close to the vehicles, but they took no notice. Through fear they tried to hide behind the tanks, they were not experienced combat troops.

On the afternoon of 28 March, it was reported to Albert Ernst who was in Jagdtiger No X1, "advancing Sherman tanks". Albert

Plate 312. Jagdtiger of 1/512. It was stopped by a close range shot which knocked its front left drive sprocket off. It burned out through a demolition charge set off by the crew who escaped into the trees. Area south east of Siegen (U.S. Army).

Plate 313. An impressive front view of Jagdtiger chassis number 305058 after being abandoned by Lt. Sepp Tarlach on 1 April 1945. The I.C is an intelligence marking (Tank Museum).

Plate 314. Front view of 305058 (Tank Museum).

Plate 315. Rear view of 305058. This was the first Jagdtiger to be fitted with the handle above the rear doors. The M.G. 42 is still on its mono-pod (Tank Museum).

Plate 316. No X7 was a 1/512 tactical number. The impact which damaged the engine is visible as well as the one which dislodged the spare track hooks (Tank Museum).

Ernst immediately gave the order "firing halt". The Jagdtigers stopped, they quickly slewed into positions, the cannons were immediately released from their travel locks and the guns loaded, Ernst's own gunner Fw Colany aimed at the lead Sherman and fired. His first round ripped the Sherman apart (this was the first kill for the Battalion).

Because the 12.8cm ammunition created a huge cloud of smoke when the cannon fired, this obscured the Jagdtigers. The militia thought that the Jagdtigers had been destroyed and ran away, never to be seen again. The Panzer Jägers then had to keep suspending firing operations; the white smoke from their own rounds was restricting their view. The Americans return fire had no effect on the Jagdtigers. They only managed to blow off some of their heavy camouflage covering.

For 3 days, 1/512 covered the retreat of the Harpe Army. They fell back through Neidernephen and Obernephen to Siegen. Throughout this period 1/512 lost 4 Jagdtigers, one had its front left drive-sprocket shot off, the other three broke down and were blown up. The unit from strategically advantageous positions had been able to pick-off, at long range, the pursuing American tanks, some being destroyed at ranges well over 3000m, this significantly retarded the Allied attacks from Herborn, over thirty vehicles (mostly tanks) had been destroyed.

By 31 March, 1/512 was back in the Siegen area with their Jagdtigers. The following day, they were involved in a firefight in Obernephen. During the seesaw battle for the village one Jagdtiger No X7 was damaged by friendly fire, it was abandoned. This vehicle (305058) was later the subject of an Allied intelligence inspection. From Siegen, 1/512 had been planned to become part of the breakout attempt to stop the Ruhr encirclement, an attack well to the north of their present position.

Because of his recent success, Albert Ernst was summoned to Battalion HQ in Siegen. He was promoted to Hauptmann, The 1/512 were enlarged to a battle group by the addition of a platoon

Plate 317. No X7 was the first Jagdtiger belonging to s.Pz.Jg.Abt 512 to be examined intact (U.S. National Archives).

Plate 318. Jagdtiger No X7, chassis No. 305058, abandoned in front of the house of the Schröder family in Obernephen. The house was demolished, in 1973, to make room for a supermarket (U.S. National Archives).

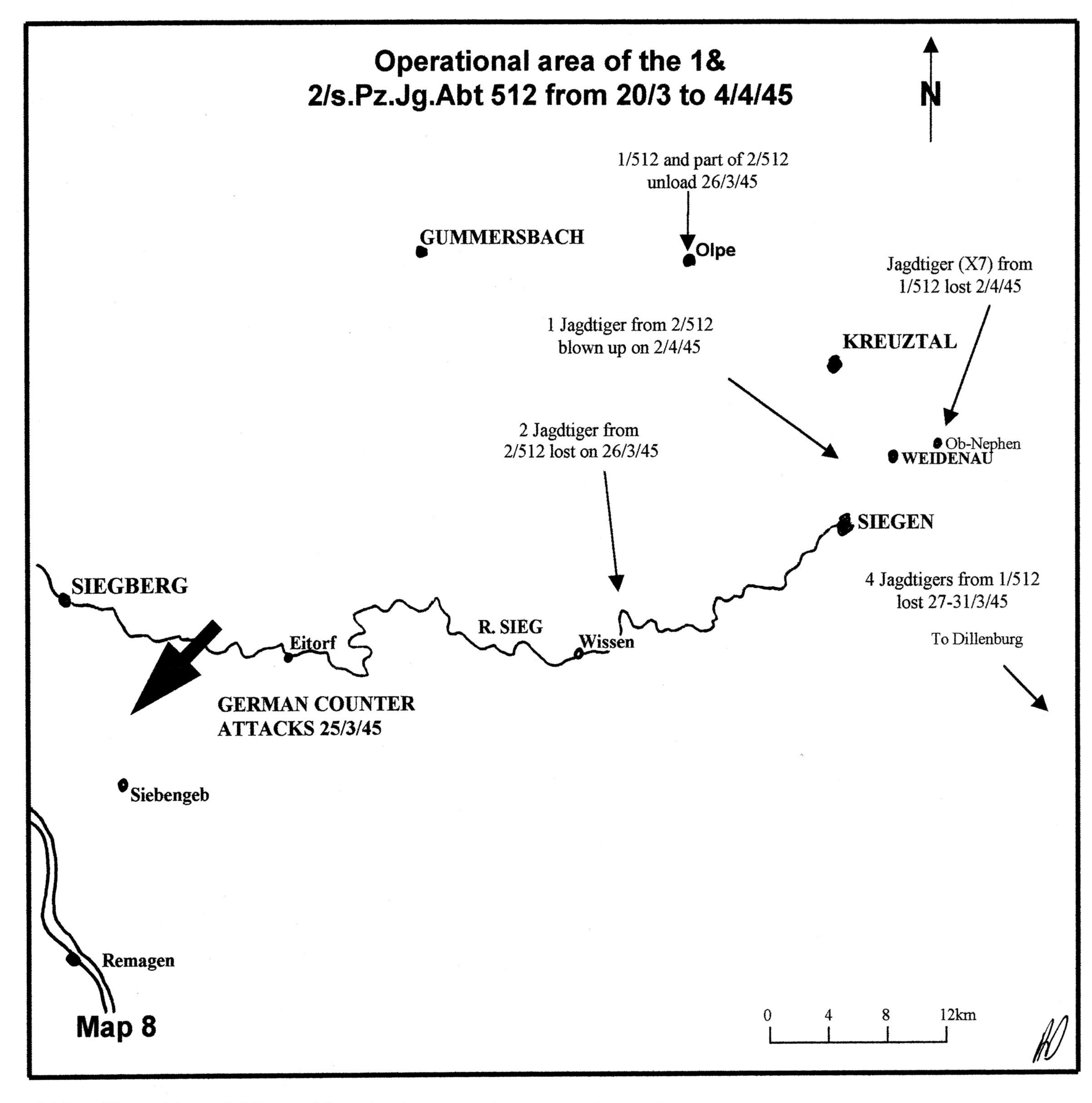

of 4 Stug III's, a platoon of 4 Panzer IV's and a platoon of 4 3.7cm flak vehicles.[7] Heinz Rondorf, who was promoted to Oberleutnant, took over commanded of the Jagdtiger Company and Jagdtiger No X1 with its extra radio set (Fu 8), there were now only five remaining Jagdtigers with 1/512.

The new structure of Battle Group Ernst was as follows:

Commander Hptm Albert Ernst
Adj. Lt. Offz Anwarter Sepp Tarlach
Special Missions Offz Oblt Boghut
Commander Jagdtiger Company Oblt Rondorf
Commander Flak Company Oblt Rudolph [8]

Battle Group Ernst were intended to become part of the breakout force, but they were never able to get into a position to assist with this attack because of the fighting with which they were involved in, in the Netphen/Sieg bridgehead area.

The breakout attempt had started earlier, on the 30 March, from the wooded hills of the Sauerland and struck through Schmallenburg [9], Oberkirchen and Neuastenberg, over 70km north west of Battle Group Ernst's position. The attack had hit at the Americans with overwhelming strength at Neuastenberg on 1 April. The Americans then withdrew and dug in around Mollseifen. A freak snowstorm hit the region that day which severely restricted progress. The attack continued but, on 2 April, it came to an abrupt halt after making only 12km of progress.

General Bayerlein, who commanded the attack, said that it had failed because his men were war weary and did not have the necessary supplies of fuel and ammunition to give effective power to the thrust eastwards. The escape route to the east of the trapped Army Group B, which included 1/512 and 2/512, was now well and truly cut.

On 4 April, Model issued orders to get all useable Jagdtigers up to Unna to help with the defense of the larger cities on the Ruhr.

38.14 The 3/Pz.Jg.Abt.512 (Schrader)

+ Fuhrer's Report - 31 March 1945

Ref: 3/s.H.Pz.Jg.Abt.512 Jagdtigers, from General Inspector Tank Troops.

The 3 company/Jagdtiger Abt 512 in Sennelager with 5 Jagdtigers, ordered by Ob. West through OKH, to go into action. Since 26.3.45, 2 further Jagdtigers are in conveyance from St. Valentin to Sennelager. 1 Jagdtiger is in St. Valentin for repair until 31.3.45.

The organization of 3/512 was as follows:

The 3 Company Commander Oberleutnant Schrader

1 Platoon Commander Lieutenant Nowak (3 Jagdtigers),
Fw Franke commanded another of these Jagdtigers.
2 Platoon Commander Obfw Anton (2 Jagdtigers),
Fw Becker commanded the other of these Jagdtigers.
3 Platoon Commander Lieutenant Erich Schröder (No Jagdtigers available).

They were to form the last of the three fighting Companies, 3/s.Pz.Jg.Abt 512. Their first Jagdtiger was completed in Nibelungen Werk, on 8 March 1945, and the next six on 10 March. The Jagdtigers were held outside the factory until rail transport was made available to ship them to Sennelager. One train became available and the first 5 were loaded and sent through, heavily camouflaged, to Sennelager where they arrived on the night of 25 March. This was the same day that the Allied attack was launched on the Ruhr area.

On arrival of their ordnance at Sennelager, the crews who had not been to the factory to see the Jagdtigers, started training. Major Scherf, 512's commander, had moved out of Battalion HQ in Westernholz on 20 March and drove down to Siegen where 1/512 and 2/512 were re-located for combat operations, this caused him to never make contact [10] with 3/512 who had not yet started training. All the troops including Schrader were more familiar with combat in tanks.

3/512 was expecting their next Jagdtigers to arrive, by 29/30 March. However, one had been damaged in the air raid on 23 March, while awaiting loading at the rail head at Nibelungen Werk, and it had to be returned back to the factory for repairs, the other two did not get through. The factory was also damaged in the same air raid and this factor would significantly delay its repair.

The U.S. southern attack made good progress and, by 28 March, threatened Kassel. The events of the war quickly caught up with 3/512 and after only five days training and gunnery practice, which was restricted because of lack of training ammunition, they were thrown into action near the Sennelager training grounds. The war had reached them!

38.15 Fighting around Paderborn and Sennelager

On 30 March 1945 the U.S. southern pincer thrust, was approaching Paderborn from the area of Marsberg.

The Allied plan had been predicted by Fieldmarshel Model, the commander of Army Group B. He sent orders to Sennelager to muster up as large a force as possible to stop the U.S. spearheads near Paderborn.

At his disposal were;

s.Pz.Abt. 507 with two companies totaling twenty-one Tigers II's and three Jagdpanthers, commanded by Hptm Schöck.

s.Pz.Abt. 508 with six Panthers and one Tiger I, commanded by Hptm Stelter.

3/s.Pz.Jg.Abt. 512 with five Jagdtigers commanded by Oblt. Schrader.

3/501 (511) with two Tiger I, one Panther, one Pz IV, all training Panzers commanded by Oblt Busch, manned with instructors.

Pz Kp Kummersdorf with 1 Jagdtiger, 2 Tiger II's, 4 Panther, 1 Pz II, all test vehicles from Waffen Pruef 6, and manned by instructors.

The report sent to the Gen Insp. d. Pz. Tr. was as follows:

Gen. Insp. d. Pz. Tr.
H. Qu. OKH, the 31.3.45
Abt. Org. Nr. F 440/45 g.K.

Fuhrer-lecture 31.3.45
List of Pz. And Pz. Gren. Unit

5.) Pz.Kp. "Kummersdorf"
From reserve Abt. Wa.A. Kummersdorf in list for ground position operation
Structure:
3 Pz. Platoons (moveable)

Plate 319. The Panzers of Pz.Kp. Kummersdorf, in late March 1945 (Delta Publications).

1 Pz. Flak Platoon
1 Gren. Begl Platoon
1 Pz. Platoon (unmoveable)

List:
3 Pz Platoons (moveable)

1 Tiger II
1 Jagdtiger[11]

4 Panther
2 Pz IV lg.
1 Pz III (5cm L 60)

1 Nashorn
1 Hummel MGK (Triple)
2 Sherman

1 Pz. Flak Platoon:
1 4 barrel Sp. Wg. (7.5cm L 24)

1 4 barrel Sp. Wg. (2cm)

1 captured Flak. Wg. (Twin)

1 B IV C (2cm)

2 B IV C (with MG)

1 Pz. Platoon (unmoveable)

1 Tiger Porsche 8.8 L/70
1 Waffenträger Steyer 8.8 L/70
1 P 40 (I)

By the morning of 30 March, Task Force Welburn led the U.S. attacking force from the south. Up to now they had experienced little resistance in the area and were moving northeast towards Burcher. A platoon of four Tigers II's from 3/507 was in the woods between Dörenhagen and Dahl.

Aerial reconnaissance by the U.S. Air Force on the line of advance spotted the four Tiger II's and a follow-up airstrike hit the German tanks with napalm bombs. They were reported as destroyed, to the U.S. ground force.

The U.S. Task Force continued to advance towards the "knocked out tanks" until the four Tiger II's opened fire at a range of just over 1000m. Seven of the first ten Shermans were destroyed; the rest quickly tried to drive for cover. The U.S. commander radioed for assistance. A second Task Force came forward from the southeast and six other Tiger II's broke cover from the woods, catching the U.S. armor in a vicious crossfire. The whole area was engulfed in thick black smoke from burning tanks. In the confusion, the U.S. General Maurice Rose was captured when he mistook a Tiger II for a U.S. Pershing, he was shot dead by mistake while surrendering to the Tiger II crew. This single action was to serve to strengthen the U.S. resolve in the whole Ruhr area.

After the ambush, the Tigers took cover in the woods near Dahl to avoid the inevitable follow up air-attack. There were no German tank losses in this action but American wrecks littered the area.

On 31 March, Easter Sunday, the Americans brought more armor into the area. Four Tiger II's and three Jagdpanthers from 2/507 stopped one attack from Eggeringhausen. Three Tiger II's were destroyed in a U.S. ambush.

The 3/507 was assembled near Schwaney ready for an attack southeast towards Warburg.

The five Jagdtigers of 3/512 were ordered to leave Sennelager to take up positions around Paderborn.

On 1 April 1945, the American attack resumed towards Lippstadt, near which, the link up of the two pincer arms was achieved at 1300 hours. The American armor had simply bypassed the German Panzer force.

The same day, a German attack was made with s.Pz.Abt. 507 moving southeast to Neuenheerse while a small Kampfgruppe of s.Pz.Abt. 508 with its six Panthers and Tiger I attacked down the road towards Lippstadt from Paderborn. The attack failed and after a withdrawal the remaining tanks took up positions near Paderborn station. The five Jagdtigers of 3/512 moved into the city of Paderborn to take up positions to cover the southwestern approaches to the city.

The 3/501 and the five Jagdtigers of 3/512 repulsed an American attack on Paderborn. One Tiger I was knocked out at Paderborn station.

On 2 April, a much stronger American attack was directed towards Paderborn and Sennelager. One Jagdtiger was damaged

Plate 320. The first Jagdtiger from 3/512 to be lost, it was blown up in Paderborn, on 2 April 1945. Note the brackets for the 2 ton crane. On the original photograph the chassis number 305068 can just be seen (Wolfgang Schneider).

Plate 321. Frontal view of a second Jagdtiger from 3/512 thought to have broken down near Paderborn. It is fitted with the crane mounting brackets (Karlheinz Münch).

Plate 322. Right side view of the same vehicle. The "Y" is clearly a tactical marking, and indicates the 3/512 company (Karlheinz Münch).

and had to be blown up by its crew, a second Jagdtiger, tactical marking "Y" broke down and had to be abandoned and blown up, the damage indicates its ammunition and fuel were removed beforehand.[12]

The remaining three Jagdtigers took up position east of the city. The s.Pz.Abt. 508 attacked the Americans at Willebadessen with nine Tiger II's. This was not successful and only five Shermans were destroyed for the loss of five Tigers II's. The remains of this Battalion withdrew to the Weser River the next planned line of defense.

The fighting continued in Sennelager for almost a week. The 3/512 with their three Jagdtigers carefully drove to the Weser River at Gieselwerder.

Plate 323. The gun still being held in the travel lock would indicate that it had not been in action when it was self-destructed (Karlheinz Münch).

Plate 324. The engine hatch complete with its monopod had been blown off (Karlheinz Münch).

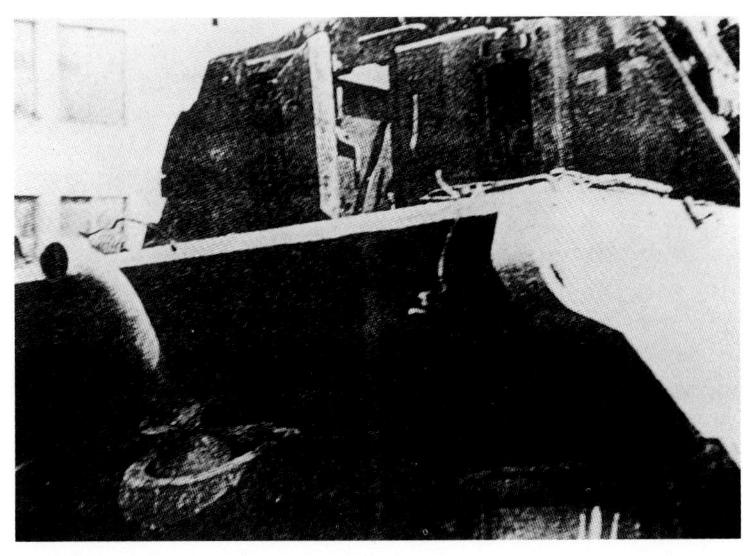

Plate 325. This rear view shows that the fighting compartment roof has been blown off. Again the "Y" marking can just be seen on the left side rear door (Karlheinz Münch).

38.16. Jagdtiger deliveries in March 1945

Delivered to	Transported on	No	Type	Chassis No	Tactical No
s.Pz.Jg.Abt.512	*03/3/45*	*5*	*(H)*	Chassis No's not in series, all mixed up due to changing steering units (305050 - 305077)[13]	
s.Pz.Jg.Abt.512	*05/3/45*	*5*	*(H)*		
s.Pz.Jg.Abt.512	*09/3/45*	*5*	*(H)*		
s.Pz.Jg.Abt.512	*10/3/45*	*5*	*(H)*		
s.Pz.Jg.Abt.512	*14/3/45*	*5*	*(H)*		
s.Pz.Jg.Abt.512	*26/3/45*	*2*	*(H)*		

Total delivered to schwere Panzer Jäger Abteilung 512 (27)

38.17 Jagdtiger built by Nibelungen Werk in March 1945

There were 3 Jagdtigers completed in March against programme of 20.

Chassis No:305075
305076 } 12.8cm Jagdtigers
305077

38.18 Hulls built in Eisen Werk in March 1945

There were 2 Hulls completed in March giving a running total of 114.

Bomb Attack on factory	20 March	18 Tons HE
" "	23 March	258 Tons HE

Plate 326. View of its breechblock (Karlheinz Münch).

[1]This statement indicates that work may have started in Ni-Werk on the 8.8cm version of Jagdtiger because of the restricted numbers of 12.8cm gun mounts available.
[2]Breslau had been surrounded by Soviet forces on 23.2.45
[3]In compliance with clause 4a of the order 11.2.45 ObdE/AHA/Staff Ia (1) Nr.9708/45 secret.
[4]The 4 x 3.7-cm flak tanks were the only ones supplied to s. H. Pz.Jg.Abt. 512, these were assigned to battle group Ernst at the beginning of April.
[5]Rumor probably created by the successes of 1/512's Jagdtigers!
[6]When specifically questioned Otto Carius stated that *all* of his Panzers got to the unloading area, for operations. However, he did state that on many occasions he was out of touch with his Panzers, this is contrary to this report. More accuracy should be attributed to official documents from the time!
[7]These were the only flak vehicles supplied to Pz.Jg.Abt 512 in accordance with KStN 1196.
[8]On 26th March, Oblt Rudolph had been ordered (Abt.Org.II. Nr. F 1340/45 geh.v.26.3.45) to transfer from s.H.Pz.Jg.Abt. 653 to s.H.Pz.Jg.Abt. 512 by Insp. Gen.Pz.Tr. (Thomale). It is not known if or when he arrived.
[9]There is an unconfirmed report that 3 Jagdtigers from 1/512 were involved in this attack. Albert Ernst was not able to confirm this before his death, however, Walter Scherf when specifically questioned on this matter, stated that no Jagdtigers from his unit were ever in this area, he also stated that the first four Jagdtigers lost by 1/512 were nearer Netphen than Dillenburg.
[10]It is interesting to note that when specifically questioned, Walter Scherf has no knowledge either as to the existence of the third company and stated only two companies with the battalion, or of any Jagdtigers fighting in the Paderborn/Harz areas.
[11]Jagdtiger 305004.
[12]This vehicle was later taken to Sennelager by the Allies see part III.
[13]One Jagdtiger damaged by bombing - later repaired. Stayed until April before it could be repaired.

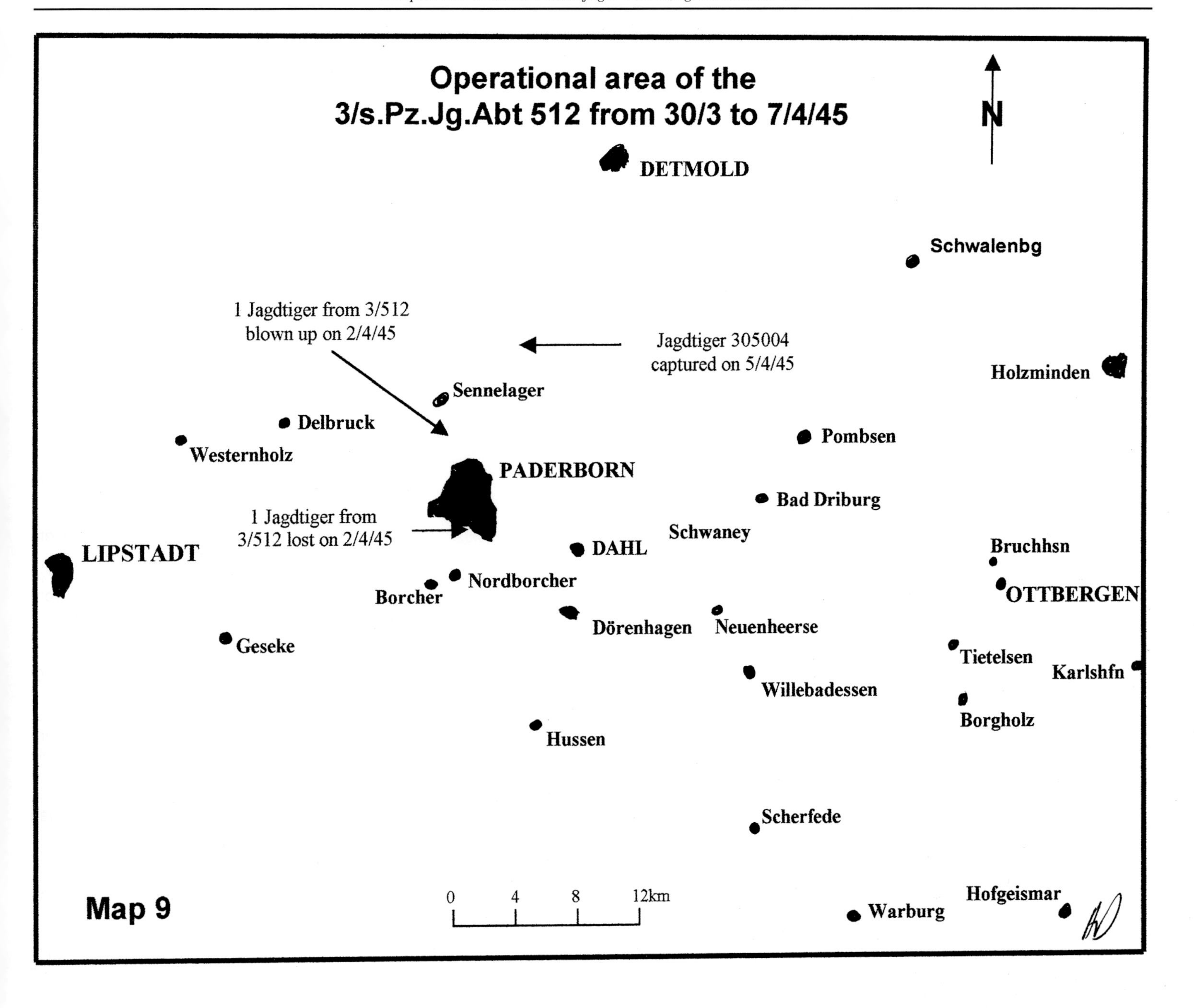
Operational area of the
3/s.Pz.Jg.Abt 512 from 30/3 to 7/4/45
N
DETMOLD
Schwalenbg
1 Jagdtiger from 3/512
blown up on 2/4/45
Jagdtiger 305004
captured on 5/4/45
Holzminden
Sennelager
Delbruck
Westernholz
Pombsen
PADERBORN
Bad Driburg
1 Jagdtiger from
3/512 lost on 2/4/45
Schwaney
LIPSTADT
DAHL
Bruchhsn
Nordborcher
Borcher
OTTBERGEN
Dörenhagen
Neuenheerse
Geseke
Tietelsen
Karlshfn
Willebadessen
Borgholz
Hussen
Scherfede
Map 9
0
4
8
12km
Warburg
Hofgeismar

39

Schwere Panzerjäger Abteilung 512 April 1945

39.1 Combat diary - April 1945

1 April The sealing of the Ruhr pocket traps Army Group B (350,000 troops). 3/512 is in action around Sennelager with 5 Jagdtigers trying to stop this. 2/512 and 1/512 are near Siegen engaging U.S. tank attacks.

2 April The 2/512 stop the capture of Siegen. 1/512 cannot get to the battle area for the attack out of Ruhr pocket. Two Jagdtigers from 3/512 lost in the Paderborn area.

4 April Fieldmarshal Model orders that all-useable Jagdtigers to deploy to Unna. 2/512 load onto railway flats that night at Gummersbach. 1/512 set off by road through Olpe, Meinerzhagen towards Altena.

7 April The 2/512 arrive at Menden in the hours of darkness. The 3/512 is in a firefight across the Weser River.

8 April The 1/512 is in Altena. 2/512 moves into position and engages U.S. tanks at Unna. The 3/512 loses 1 Jagdtiger near Offensen.

9 April Unna falls. The 2/512 moves south towards Langschede. One Jagdtiger lost to enemy action. The 1/512 drives through Hagen to Ergste, for a refit.

11 April 1/512 and 2/512 both engage U.S. attack near Langschede. The 512 ordered to fall back south of the River Ruhr. 3/512 lose a Jagdtiger at Biersfelde.

12 April 2/512 is in Kalthof. The 1/512 guards the airfield at Deilinghofen.

13 April Menden falls and the village of Kalthof is surrendered.

14 April 1/512, Ernst hands over Hemer and moves to Iserlohn. 2/512 moves to Letmathe-Ergste. The 4 Jagdtigers at Linz are assigned to s.Pz.Jg.Abt.653.

15 April 1/512 is in action around Iserlohn. 2/512 is having refit near Ergste. The 3/512 is in action with 1 Jagdtiger near Kamschlacken. 2/512 surrender, their last 6 Jagdtigers are self-destructed to prevent hand over.

16 April 1/512 surrender Iserlohn with full military honors. The last 3 Jagdtigers are handed over to the Americans. Walter Scherf and Fritz Bayerlein surrender.

18 April The battle of the Ruhr pocket is over.

21 April Walter Model, Commander Army Group B, shoots himself.

22 April The battle of the Harz pocket is over.

39.2 Fighting around Siegen

In the Southeast of the Ruhr pocket lay the small town of Siegen, a key road and railway center.

While the U.S. 1st Army was fighting up around Paderborn, in their rear the U.S. 15th Army was trying to cross the River Sieg with the key town of Siegen as their first objective.

On the first day of April, in spite of poor morale, the Germans fought desperately, launching 17 counter attacks against the Americans. The Americans lost one of their two bridgeheads over the River Sieg north east of the town.

On 2 April, the American 8th Division hit considerably more resistance while trying to take the key heights in that sector that it needed before it could proceed any further. Walter Scherf was in the area with Otto Carius, Albert Ernst and both 2/512 and 1/512.

The Jagdtigers were placed on high ground. The 2/512 receives orders to stop the American troops coming from Dillenburg, especially the tanks and other vehicles.

Scherf was himself with a Jagdtiger. A second Jagdtiger also from 2/512 was on high ground on the north side of Siegen. Below him, was the village of Eiserfeld. They were shooting over Siegen at the American vehicles moving north. The Supply Company of Oblt Fink was on the western zone outside Geiswald, together with the Workshop Company who were still working on 2 Jagdtigers from 2/512. The rest of 2/512 with 4 Jagdtigers had to relocate frequently.

In the area of Siegen the Jagdtigers went into positions on the hill near the barracks. Battalion HQ was still there, however, there was plans to move it to a safer area. The Americans were alerted and halted.

The next assignment for the Jagdtigers was the securing of an anti-tank barrier near Weidenau. Four Jagdtigers made an attack due east of Weidenau to recover a hill from which the Americans threatened the German positions.

The following morning 4 Jagdtigers drove south at high speed straight towards the American position, they fled in panic with the hill being quickly taken. From the hill Oblt Carius spotted an American tank that disappeared behind a house for cover - his Jagdtiger aimed at the house and fired. After 2 shots the U.S. tank burned out completely. This was the first kill for 2/512.

The Americans resorted to their usual tactic and directed artillery fire and fighter-bombers at the Jagdtigers' positions. Withdrawal of the Jagdtigers into the woods occurred without loss. No, German infantry appeared to take over the captured hill. As it would have been stupid staying in this position, the 4 Jagdtigers withdrew, before the inevitable follow up air-attack. During this withdrawal, one Jagdtiger became bogged down in a bomb crater. It was blown up after repeated attempts to pull it out failed.

Shortly after, 2/512 received orders directly from field Marshall Model, to proceed north to Unna to help with the defense of the city.

The American 8th Division, (VII Corps), took 48 hours to capture the key heights that it needed before it could proceed any farther with its attack on Siegen.

On 4 April, Walter Scherf received the order from Field Marshall Model: "all useable Jagdtigers to be brought back on rail transport to the area of Frondenberg/Langschede to move to Unna."

39.3 Supply Problems

Oblt Hans Fink was in charge of the Supply Company of 512 and was responsible for all supplies to the Jagdtiger Battalion. His biggest difficulty was getting the gasoline to keep the vehicles operational. He was always looking for gasoline. The Jagdtigers had very thirsty engines. He found out the existence of a tanker train loaded with gasoline, hidden in a tunnel near Werdohl. He accompanied Oblt Carius and the transport vehicles with as many of the 100-liter fuel drums as could be collected.

On arrival at the tunnel they ordered the train driver to drive his train out of the tunnel, they filled everything up, but in the haste there was a lot of spillage. The train driver asked for their names. Later, there was a lot of trouble because the tanker train was the OKW reserve. They did not know this at the time!

This initiative taken by Hans Fink and Otto Carius kept the Jagdtigers and transport vehicles of 512 operational for the last 10 days of their short combat period.

39.4 The 2/s.Pz.Jg.Abt.512 sent into action at Unna

As previously stated on 4 April 1945, orders redeployed 2/512 to move their Jagdtigers to Unna. The Jagdtigers were in the process of being loaded onto rail flat cars at Gummersbach when the U.S. broke through the German lines at Weidenau making it difficult for the Jagdtigers to get under way. Fighter-bombers were attacking the railway network on a continuous basis that resulted in the rail personnel refusing to take responsibility for the journey. To retrieve the situation 2/512 own troops staffed the locomotives themselves. They sent a small shunting engine in advance to assess the condition of the rail line. The company commander went quickly by road to Unna to assess the terrain for operational suitability. On arrival, he found that the Americans had already taken Werl and this limited their area of deployment. The Jagdtigers unloaded at Menden and were then driven by road to Unna.

At Unna 2/512 with it's 7 Jagdtigers were attached to a local defense command. The 7 Jagdtigers were deployed, 2 covering the Ruhr highway pointing east towards Werl, the other 5 were hidden in the northern outskirts of Unna facing north towards Kamen, the direction from which they expected the first American attack.

The following day, 8 April 1945, the Americans were firing into the town with tank fire from a great distance. The Americans started to crawl forwards. Their vehicles were rolling quietly past Unna towards Dortmund, unaware of any German presence in the city. Soon afterwards the 2 Jagdtigers hidden in a cemetery that covered the Ruhr highway opened up on the American column. Vehicles caught fire and the rest of the traffic immediately broke up and drove wildly back eastwards. With over 20 tanks and armored cars destroyed this was 2/512's biggest success. The American's fear delayed the fall of Unna for a further day.

Just before daybreak on 9 April, Unna was rapidly becoming encircled. The 7 Jagdtigers withdrew from Unna and drove south on the B233 highway to the next village. Five American tanks started to move south along the B233 towards the Jagdtigers. Oblt Carius deployed a single Jagdtiger to stop the Americans. A young tank commander, who had no experience at the Front, wanted to handle the matter himself. Carius personally escorted the Jagdtiger to a firing position so nothing could go wrong. The commander got into his vehicle while Carius stopped to watch, he made a fatal mistake and did not release the gun from its travel-lock until he arrived on the high ground. The Americans heard his engine, 2 tanks disappeared, but the other 3 fired at the Jagdtiger and soon hit its front armor. Instead of firing, this commander decided to turn his tank rather than reverse it backwards, giving the Americans a broadside target. The Jagdtiger went up in flames, killing all 6 crew. This was another example where even the best weapon in the world is useless without expertise and experience.

The 2/512 was now down to 6 operational Jagdtigers. An assault with 3 Jagdtigers followed, destroying one of the Shermans. The rest drove quickly back towards Unna. The Americans did not push any farther south that day as they directed their efforts west towards Dortmund. Unna fell on the 9 April 1945.

39.5 Fighting east of Paderborn

By the 2 April, part of Paderborn and the SS training camp at Sennelager had become a salient in the Allied front line, bitter fighting continuing in the area. The German defending force named SS Panzerbrigade West Falen, comprised the remnants of s.Pz.Abt 507 with eighteen remaining Tiger II's and three Jagdpanthers, plus troops from s.Pz.Abt 508 and 3/s.Pz.Jg.Abt 512 with three Jagdtigers.

Over the next four days the Panzer force were to split into small battle groups to hold positions east of the Weser River. Fighting took place around Willebadessen, on 2 April, with the loss of five Tiger II's for five Shermans destroyed. On 3 April, one Tiger II broke down in Pombzen. By 5 April, the fighting was in Ottbergen and Potzen with the loss of three Tiger II's.

Plate 327. Photo taken on 6 April 1945. Jagdtiger chassis No 305004 was captured in Sennelager by the Americans (Tank Museum).

Plate 328. A posed shot of the Jagdtiger and its new owners (Tank Museum).

By 6 April, the fighting was in Schwalenberg in the North and Tietelsen in the South, that night the three Jagdtigers of 3/512 withdrew across the Weser River. That same day the fighting in Sennelager was virtually over and the Waffen Pruef 6 Jagdtiger captured. A photographer from the Keystone Press Agency almost immediately photographed it.

Plate 329. The chassis number and Waffen Pruef 6 number – 305004 and 253 are still clearly visible in this photograph (Tank Museum).

On the 7 April, 3/512 were in the Weser River area firing at American targets across the Weser River near Gieselwerder, while parts of s.Pz.Abt 507 (one Tiger II and one Jagdpanther) were near Karlshofen. They destroyed 17 U.S. tanks in the crossfire. Friendly machine gun fire killed the company commander Oblt Schrader. Lt. Erich Schröder took command.

By 8 April, a U.S. attack was directed west towards Göttingen and Wahlsburg. This was stopped at Adelebsen by the last Jagdtiger of I Zug 3/512 and Tigers II's of 507. The German force disengaged to avoid air attack by driving into woods for cover.

The Kampfgruppe withdrew through Offensen where the last Jagdtiger from I Zug threw a track at speed and slid straight off the road and down the embankment, the crew were all shaken from its sudden unexpected stop. There was no chance of recovering it; other than by using methods, which would risk damage to the other Panzers, neither did they know what mechanical damage might have been done. It was ordered to be blown after its radios, machine guns, optical equipment and ammunition were removed, even some of its gasoline was siphoned off. 3/512 now had only two Jagdtigers remaining and both were operational.

The Jagdtigers were loaded onto the train at Vierlihausen and traveled that night through Göttingen, which was under threat from the American spear head units, it was very lucky to get through!

The withdrawal route for s.Pz.Abt.507 was through the forest roads from Ulsar via Bollensen, Hardegsen to the Harste - Parensen area, where they were in action against the Americans on 9 April. 4 Tiger II's were lost in the second assault. The last 2 Tiger II's drove across the Leine River over the Road Bridge at Nörten Hardenberg, which they held open, for troop and equipment withdrawal. The train carrying 3/512 unloaded in Nörten Hardenberg, the Jagdtigers were equipped only with transport tracks.

Plate 330. Side view of Jagdtiger from 3/512 photographed near Offensen, on 9 April 1945. Its left track broke, causing it to slew off the road (Delta Publications).

Plate 331. A very close examination of this photograph reveals part of the chassis number 305074. The vehicle is fitted with the mount for the star antenna, and also has mounts for the 2 ton crane (U.S. National Archives).

39.6 The 1/s.Pz.Jg.Abt.512 cross pocket to assist in northern breakout attempt

The new Battle Group was ordered north to Altena. They went by road from Siegen through Meinerzhagen, Bruggen and Ludenscheid a road march of 40km. They arrived at Altena throughout 7 April 1945.

On 8 April, 1/512 received the orders to break out of the Ruhr pocket at Unna. The Jagdtiger Company along with the Works Company, drove north through the outskirts of Hagen and camped in woods between Ergste and Burenbruch on the 9 and 10 April where, repairs and maintenance was carried out on the Jagdtigers. The other armored vehicles drove through Iserlohn to Langschede, followed by several battalions of Panzer Grenadiers. One Jagdtiger had to be left near Hagen, because of severe mechanical problems. It was later plundered for some spare parts and blown up on 11 April.

Plate 332. Front view of the remains of Jagdtiger chassis number 305057 photographed near Hagen by Peter Gudgin, in June 1945 (Peter Gudgin).

Plate 333. A close look at the photograph reveals the number X5. This was assigned to 1/512, and was originally commanded by Lt. Kubelka. Peter stated from his notes that there was a destroyed Hummel in close proximity (Peter Gudgin).

The next morning, 11 April, the Jagdtigers moved north along the B233 to Langschede where they crossed the Ruhr River. Here the whole Battle Group assembled ready to proceed north to help relieve Unna, which had fallen on 9 April. Shortly after daybreak a report was sent from a command armored car positioned 5km north on a hill near the Bismarck Tower, stating "American tanks approaching". The Kampfgruppe were ordered to deploy on the hill. From this position they looked down and saw a huge column of American vehicles moving west along the B1 road towards Dortmund. At the same time, a large group of vehicles had turned south and were driving along the B233 straight towards their position.

All the armored vehicles were deployed in positions just behind the crest. In all, there were 4 Jagdtigers, 3 Panzer IV's and 4 Stug III's, together with the 4 flak vehicles on the east side of the road, the latter were moved into positions further to the sides to cover for air attack. To the West Side and further back were the 6 Jagdtigers of 3/512.

The Americans continued moving forward down the road with their tanks in front. They were planning to drive from the north right into the middle of the Ruhr pocket. Because most of the German soldiers were just giving up this lulled the Americans into a false sense of security. The order was given to hold fire until the Americans were within range of all the vehicles. All the vehicles selected different targets and took aim. Hptm Ernst gave the order to open fire, which they did immediately. There was the sudden thunderous clatter, then they could see the smoke trails as the shells screamed towards the American armor. Below them the American vehicles suddenly started to burn, throwing smoke everywhere. The Jagdtigers, were destroying the tanks furthest away at up to 4km range, again the Jagdtigers large ammunition generated huge smoke clouds revealing their position.

The Americans in their panic to escape drove everywhere, becoming stuck some overturning. The Battle Group continued firing at this chaos. The Americans were unable to return fire. Finally, the Americans did their usual tactic and withdrew out of range of the German guns to wait for their air force to tip the scales in their favor.

Aircraft engines were heard and flares were fired to show the alarm. The 4 flak vehicles opened up on the aircraft as they came in from the north and northwest at an altitude of about 500m, the camouflaged flak vehicles opened fire and caught the airmen unawares. The lead aircraft blew apart, throwing wreckage everywhere; a second was hit which crashed 100m behind the German position. The remaining Thunderbolts flew straight over and disappeared. Within a few seconds they were back, again from the northwest, but much lower this time, 5 aircraft fired rockets which exploded against the slope, 1 completely destroyed a flak vehicle killing all the crew. The 3 other flak guns continued to fire at sortie after sortie until all were out of ammunition.

A further air attack started shortly afterwards. The Battle Group was only able to counter with machine guns. One Jagdtiger was hit by a rocket, which went straight through the back hatch. Lt. Kubelka and his crew were killed. Another Jagdtiger sustained a damaged exhaust pipe. There was now no way of defending against this type of attack and therefore withdrawal from the hill top position became necessary. The Americans had achieved their objective.

The Battle Group withdrew with the 3 Jagdtigers ordered to act as rearguard. However, the Americans were in no position to follow, their ground forces were burning on the road. The Americans had lost at least 50 vehicles, including 11 Sherman tanks. The German's final attempt to breakout to the north had to be called off; the American forces were too strong.

This action on 11 April, was the biggest success for the Jagdtigers of s.Pz.Jg.Abt. 512.

Radio signal 10 April 1645

Panzer log s.Pz.Jg.Abt 512:

Inventory - 19 Panzers,(1) 14 ready for action.

39.7 The last week in combat for 2/s.Pz.Jg.Abt 512

From his position just north of Langschede, Otto Carius also watched the collapse of Unna along with the peaceful advance of lengthy columns of American vehicles driving west into Dortmund. White flags were waving everywhere; not a shot was being fired!

There was no contact with the Americans that day, the 10 April. The 2/512 was occupying the high ground west of the B233 highway just north of Langschede and the Ruhr River. The company was covering the high ground above Strickgordicke as part of the wide tank front to block the American's advance into the Ruhr pocket. The order was received to hold at all costs for as long as possible, because the Ruhr pocket would collapse from this critical point.

On 11 April, the Americans started moving south along the B233 towards the tank front. They did not expect any resistance and their reconnaissance did not detect the German presence in the area. These 6 Jagdtigers along with the tanks of Battle Group Ernst hit the American column with such ferocity at a very long range that the American armor stopped and retreated. The 2/512 Jagdtigers had been firing HE shells at over 5km along a clear sight line; Major Scherf had coordinated the ambush.

That evening 11 April, the order came from Lt. Gen Bayerlein to disengage and take up new positions on the southern side of the River Ruhr, after which the bridge was blown up.

Major Scherf knew the Americans never conducted offensive actions at night. Throughout the night 2/512 were withdrawn down the B233 to a new position just north of the village of Kalthof, the Jagdtigers being camouflaged on the high ground, from where they had a good line of fire up the B233 towards Langschede.

On 12 April, the Americans again moved forward south along the B233 towards the Ruhr River. The 2/512 were the only remaining tank unit in the area as 1/512 had been moved southeast to Hemer. When the Americans got within range (3km) the Jagdtigers hit them again, and stopped their advance. The firing caused protest by the head doctor from the hospital in the village. It was overflowing with casualties. A surrender of the village was negotiated. It was laid out exactly how the Americans would occupy the village while the Germans left, with the time of the truce being set precisely.

The Jagdtigers drove southwest out of the village to Leckingsen, and then northwest to Ergste. After 2 days, 2/Company ended up in the village of Ergste. There was no fight left in any of the troops. The Jagdtigers were hidden in a patch of woods near Letmathe with the Works Company busy carrying out repairs and maintenance.

On 15 April 1945, word was sent to 1/512 that 2/512 were in the area of Ergste. At this time, 1/512 was in Iserlohn and sur-

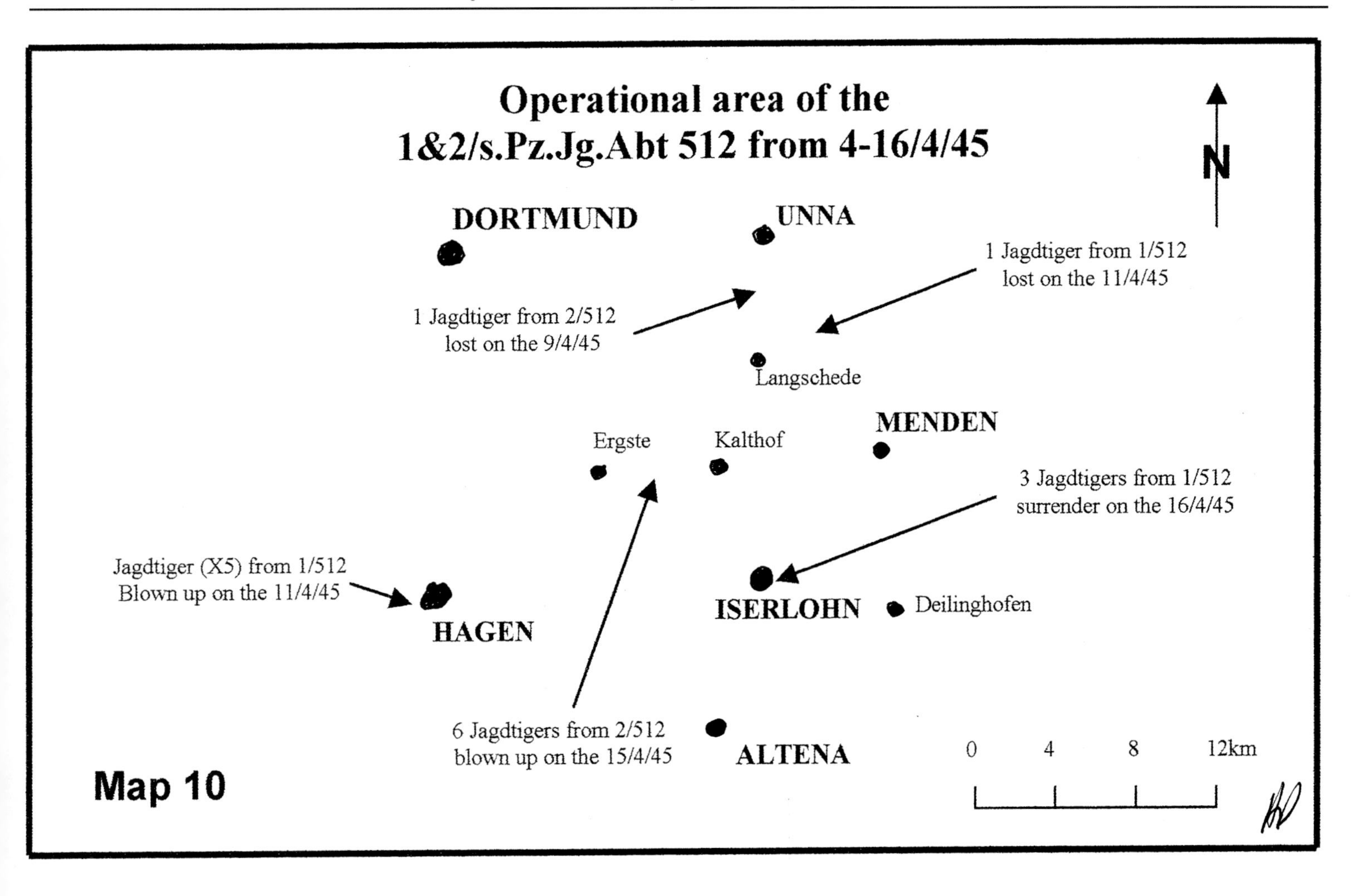

rounded with only 3 Jagdtigers remaining. That day, news came to Oblt Carius that the Americans were in Ergste, the situation was deteriorating rapidly, no troops other than his own could be relied upon, even his own battle hardened men were also ready to give up the fight. That day, Otto Carius gave the order to blow up the last 6 Jagdtigers, the company assembled for the last time, and good-byes were said. So ended the 2/s.Pz.Jg.Abt.512, on 15 April 1945.

39.8 The Last Days of 1/s.Pz.Jg.Abt. 512

After disengaging from combat with the Americans on 11 April 1945, the German force was withdrawn slowly from the position. This had to be done, as no anti aircraft ammunition was available to counter further air attacks. The Battle Group began its withdrawal while the 3 Jagdtigers acted as rearguard. One of the 3 Jagdtigers had been slightly damaged by the air attack, but was still operational. Ernst watched the American land forces from his

Plate 334. Two Jagdtigers from 2/512 parked on the edge of a wood near Letmathe – they are facing in opposite directions. The one in the background had probably towed the other into this position. Both have been self destructed (U.S. Army).

Plate 335. A British soldier and a little girl stand on the remains of a Jagdtiger. The fact that its gun has been sabotaged, along with it being in the British sector of occupied Germany leads one to conclude it was also from 2/512 in the Letmathe area.

Plate 336. Two Jagdtigers of 1/512 in Iserlohn (U.S. Army).

hilltop position. They had incurred great losses and were not in a position to pursue the German withdrawal. 40 vehicles and 11 Sherman tanks burned throughout the day.

On the morning of the 12 April, after having crossed the River Ruhr, the bridge was blown up, 1/512 received orders from Lt. Gen Bayerlein to proceed to Deilinghofen and defend the airfield for 24 hours. Ernst led his Battle Group to Hemer. On route he saw the local hierarchy proceeding to the airfield, leaving the sinking ship. Hptm Ernst used a house in Hemer as a Command Post. The Jagdtigers, tanks and anti-aircraft guns set up a defensive position around the airfield. Fortunately, they had been able to scrounge up some ammunition for the anti aircraft guns. The armor was formed into small battle groups and placed on local high ground. One Jagdtiger was in a rope factory near Bilveringsen; the other 2 were on a ridge at Hemerberge. Grenadiers were securing the airfield. Anti aircraft guns already located at the airfield became part of Battle Group Ernst.

In the evening the Americans began moving forwards, 2 Shermans were destroyed at very long range, over 4000m, by the Jagdtigers. No further American advance occurred that night.

All was quiet on the beginning of 13 April. Menden, a few kilometers to the north of Hemer, had fallen. Ernst knew the end was near. Shortly, the 2 Jagdtigers on the Stockschlade ridge fired on the Americans coming from the north, 2 American tanks were destroyed, and the others quickly withdrew. Hemer came under artillery fire. A Medical Captain from the hospital in Hemer came to see Ernst to see if they would cease fighting near the hospital. Surrender was effected. All German units went into captivity except Battle Group Ernst, which was withdrawn, with all its weapons during the time of the truce. Ernst advised the Americans to drive to Iserlohn under a white flag to see about surrendering the whole of the Ruhr pocket.

On 14 April, Hptm Ernst and his Battle Group withdrew to Iserlohn, which was already burning from artillery fire. There he reported to Lt. Gen Bayerlein who ordered Ernst to report to General Buchs in Iserlohn. The heavy armor took up positions covering the arterial roads approaching Iserlohn. Ernst received a message that 3/512 had arrived in Ergste.

Ernst reported to General Buchs, who was furious about the hand-over of Hemer. He immediately ordered Ernst to make a reconnaissance of the area. Ernst made his reconnaissance, which revealed that only his troops were conducting fanatical resistance; the others were seeking refuge.

The situation was quiet throughout the night of the 14 and 15 April. In the early morning of the 15 April, the Americans started shelling Iserlohn. Ernst drove back to General Buchs and was annoyed to find that no decision had been made to contact the Americans. The city was surrounded. The General was not up to the situation and had very little idea as to what was going on. Ernst was playing with his own life when talking of surrender. He was placed under guard while the General decided his fate. On his initiative, Ernst escaped and got back into his armored car. He expected to be shot while escaping. On his return to base, Ernst informed his men what had happened and they begged him not to return to the General.

However, Ernst did return to the General, and to be on the safe side he took 2 Jagdtigers and a Stug III with him! He told his

Plate 337. Jagdtiger commanded by Heinz Rondorf moves through Iserlohn for the surrender ceremony (U.S. Army).

Plate 338. Sepp Tarlach's Jagdtiger reverses into the town square (U.S. Army).

Plate 339. A nurse and other civilians wave at the Panzerjägers (U.S. Army).

Plate 340. Rondorf's Jagdtiger reverses into the town square. A careful look at the Jagdtiger in the background reveals No X1, which is fitted with the star antenna (U.S. Army).

troops to stay alert, turn off the engines, if you hear a shot open up with the Jagdtiger cannons, then move in. On his return, Buchs had fled. A search was made, but he could not be found. Ernst then told them that he was now in command of Iserlohn.

Ernst talked to all his men and they agreed that surrender was the only answer as nearly 300,000 people were crammed into a small area. It was agreed that Ernst should personally drive to the Americans under a white flag.

Throughout the morning of the 16 April, the Americans continued shelling Iserlohn. Two of Ernst's men were killed; they were the last of his men to be killed in action.

Ernst drove to the American positions and eventually managed to negotiate a surrender, he drove to the City Hall, escorted by his last 3 Jagdtigers.

Ernst sat down at the negotiating table with Lt. Col. Kriz of the U.S. 99th Infantry Division. The surrender was to take place at the Schillerplatz in the presence of the entire Battle Group.

When it was over the Jagdtigers were driven into the Schillerstrasse, followed by a huge crowd of civilians. The crews got out from their vehicles and assembled in front of the Jagdtigers. Ernst turned to the American Commander and stated "the last fighters of the Ruhr pocket, the fighters of Iserlohn, surrender in the face of a hopeless situation and request honorable treatment". He then handed his pistol to Lt. Col. Kriz, who saluted Ernst and took the weapon. Ernst then turned to his men and told them to lay down their own arms and dismissed them. The Americans filmed the event.

Iserlohn was the only City in the Ruhr pocket to be honorably surrendered to the enemy.

So ended the 1/s.Pz.Jg.Abt. 512 on the 16 April 1945.

39.9 Final battle for 3/s.Pz.Jg.Abt.512

On the 26 March 1945, two Jagdtigers were sent by train from Nibelungen Werk destined for Sennelager, a third Jagdtiger which was also ready was damaged in the air attack that day and required repairs before it could be further released for combat. This Allied air raid directed at Nibelungen Werk stopped production for the next two weeks thus no more Jagdtigers would be available for 3/512, This train never got through!

Plate 341. Close up of No X1 with star antenna and attached white flag (U.S. National Archives).

Plate 342. The crew remove personal equipment from their Jagdtiger (U.S. Army).

Plate 344. Left side view of the three Jagdtiger's No X1 can just be seen on the nearest vehicle (U.S. Army).

Plate 345. The troops on parade. Note the Berge Panther in the background (U.S. Army).

On 10 April, 3/512's last two Jagdtigers were driven from Norten Hardenberg station, they were accompanying s.Pz.Abt.507 with only 2 Tiger II's, the withdrawal route was through Kaltenburg to Dorste, there had been some sporadic rearguard firing at the pursuing Americans. The 3/512 were deployed 3km south east in Schwiegershausen to block the expected American advance.

The small Panzer group was incorporated into the 11th Army as part of the plan to turn the Harz Mountains into a fortress, in total 88,000 troops with very little armor. There was, however, a large number of anti-air craft guns of all types in the Harz area, which could be used against ground targets.

The Jagdtigers were used on 11 April to stop the Americans at Schwiegershausen,[2] over 10 M4 Shermans were destroyed, Jagdtiger 321 took a hit on its front glacis plate which simply bounced off. The Americans broke off the attack and waited for air support, Schröder pulled the Jagdtigers out and placed them in the edge of a wood between Schwiegershausen and Beierfelde, still with a commanding view of the approach road. The Tiger II's were also in action against this assault at Dorste.

The following day as the Americans resumed their advance, Jagdtiger 322 (Fw Becker) shot off the last of its ammunition, he was forced to withdraw, the Jagdtiger broke a track and as there were no spares available, it was blown up.

This put the pressure onto Jagdtiger 321, which then developed a fault in its recoil cylinder, it required refilling. The Jagdtiger and the two Tiger II's withdrew to Osterode, where they were put under the control of Regiment Holzer, one Tiger II broke down and was abandoned to the Americans that day, all 5 crew were subsequently killed!

Regiment Holzer held the valley approach from Osterode leading along the B498 toward Altenau. On 15 April, 321[3] was ordered into a pine forest near Kamschlacken, it broke its left track whilst trying to reverse into the wood for better cover, a broken swing arm had caused this. The transport tracks had never been intended for combat maneuvers. The crew was able to escape into the wood. This last Jagdtiger was later shot up by the Americans who managed to dislodge its gun mantle and burn it out, it was photographed by the Americans on 16 April.

Plate 343. The last three Jagdtigers of 1/s.Pz.Jg.Abt 512 (U.S. Army).

Plate 346. A very impressive view of the last act of 1/ s.Pz.Jg.Abt 512. The middle Jagdtiger has five kill rings around its barrel (U.S. Army).

Plate 347. Front view of the last Jagdtiger from 3/ s.Pz.Jg.Abt 512. It had been pressed into action on transport tracks, one of which broke while it was trying to slew into this firing position. It sustained a hit on its front glacis plate, and a second hit dislodged its gun mantle. It was later burned out (U.S. National Archives).

Plate 348. Rear view of the same vehicle. It had been fitted with the mount for the star Aerial; vehicle in almost total red primer color (U.S. National Archives).

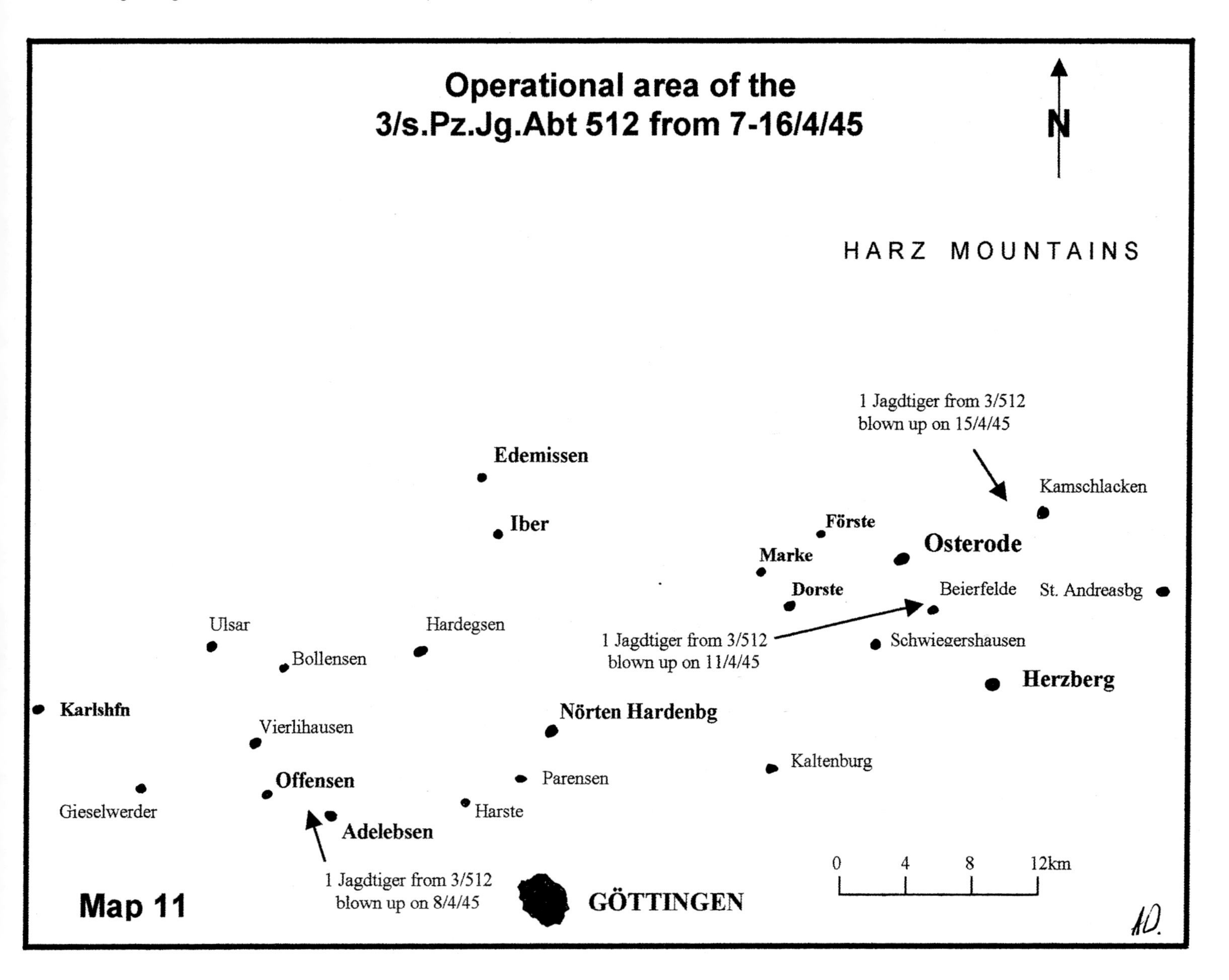

So ended the combat operations of the last Jagdtiger from 3/ s.Pz.Jg.Abt 512 on 15 April 1945. The battle for the Harz pocket lasted until 22 April.

39.10 s.Pz.Jg.Abt 512 Jagdtiger losses - March/April 1945

Second company

25.03.45	10 Jagdtigers, 8 operational, 2 short-term repair.
27.03.45	2 Jagdtigers blown up by crews near Kirchen.
01.04.45	8 Jagdtigers, 4 operational, 4 short-term repair.
02.04.45	1 Jagdtiger lost near Weidenau (stuck in bomb crater).
08.04.45	7 Jagdtigers all operational.
09.04.45	1 Jagdtiger lost near Unna (Enemy action).
10.04.45	6 Jagdtigers all operational
15.04.45	6 Jagdtigers blown up by crews near Ergste.

First company

26.03.45	10 Jagdtigers all operational unload in Olpe area.
28.03.45	4 Jagdtigers lost covering retreat south of Siegen.
01.04.45	6 Jagdtigers all operational 1 Jagdtiger lost in Obernephen (Friendly fire).
10.04.45	5 Jagdtigers, 4 operational, 1 long-term repair.
11.04.45	1 Jagdtiger blown up by crew at Hagen.
11.04.45	4 Jagdtigers operational. 1 Jagdtiger blew up by air attack near Langschede.
16.04.45	3 Jagdtigers surrendered in Iserlohn.

Third company

30.03.45	5 Jagdtigers, 5 operational, 2 more Jagdtigers in rail transit since 26.03.
02.04.45	2 Jagdtigers lost near Paderborn.
08.04.45	3 Jagdtigers operational. 1 Jagdtiger lost near Offensen.
11.04.45	1 Jagdtiger lost near Beierfelde.
15.04.45	1 Jagdtigers lost near Kamschlacken in Harz Mountains.
?? 04.45	2 Jagdtigers in rail transit did not get through, returned St. Valentin?

39.11 Conclusion

Schwere Panzer Jäger Abteilung 512 was the smaller of the two Jagdtiger units. However, they achieved a greater reliability of mobility because of the new steering units.

Again only about 20% of their Jagdtigers were lost to enemy action, the rest being blown up by the crews or surrendered to the Americans.

Four of their Jagdtigers were captured intact, all belonging to 1/512. None survive today.

Unfortunately very few German photographs taken inside the Ruhr pocket survived the war as the Allied troops confiscated the German P.O.W's cameras. Otto Carius states that there was no photographic equipment available to 2/512.

The three-company structure was maintained until the Battalion's surrender. Otto Carius has no recollection of any physical contact with Albert Ernst during his time with 512, or any knowledge of Oblt Schrader and 3 company. Walter Scherf also has no recollection of Oblt Schrader and 3 company.

When referring to the original report documents, the company numbering system (1 - 2 - 3 company) is to say the least contradictory. To save confusion I have made alterations to keep uniformity in the text, there is still some debate as to whether it was Carius 2Kp and Schrader 3Kp or vice versa, Otto Carius supported by Walter Scherf insist his was 2Kp.

The Battalion destroyed over 150 Allied tanks and vehicles.

Unfortunately I have found no official reports concerning s.Pz.Jg.Abt. 512 in April 1945 other than the Panzer log on the 10 April 1945, to be able to further substantiate the veteran's accounts!

(1) This figure must have included the 4 Stug III's and 4 Pz IV's, there were only 11 Jagdtigers still available to 1& 2/512, no Jagdtigers from 3/512 will have been included in Major Scherf's report!

(2) Erich Schröder names the place as Spiekershausen in his account, however this location is 2km east of Kassel and was over-run by the Americans on 4th April.

(3) Erich Schröder has referred to these Jagdtigers in his account by tactical numbers, however the photographs indicate that tactical numbers were not actually painted onto the Jagdtigers!

40

Jagdtigers Not with the Main Combat Units

40.1 General

When trying to trace the final usage, of the training and later constructed Jagdtigers, there is very little documented evidence available that would prove beyond reasonable doubt what really did happen and how many were deployed?

The following sections present the documented evidence along with various veterans' accounts. Unfortunately one has to resort to speculation as to the actual numbers involved! The Jagdtiger with Wa-Pruef 6 is covered in chapter 38/15 and 39/5.

40.2 Jagdtigers in Putlos Gunnery School

Putlos Gunnery School was an outstation of Training Battalion 500 and situated on a peninsula on Germany's northeast Baltic coast.

The first Jagdtiger 305005 (P), to be transported to the school, it was dispatched from Nibelungen Werk, on 14 October 1944.

Later, after s.Pz.Abt 653 had reached its full compliment of Jagdtigers, a second Jagdtiger was sent to the school on 25 January 1945, from the record this was identified as a Henschel vehicle. It was likely to have been 305049 (H).

Training vehicles were considered for combat use as the war progressed and the number of Panzers available to the Germans decreased.

On 5 March 1945, an inventory of the Panzers available to Battle Group North:

Pz Kp: with 13 Pz V and 1 Pz V1 (Tiger 1)

Light Pz Kp: with 15 Pz IV and PanzerJäger.

(There is no mention of the Jagdtigers in this inventory and if compared to the report in Chapter 38.15, which shows that even immobile vehicles were being considered for action at this time!)

On 19 April 1945, the s.Pz.Abt 510 was ordered to move to Putlos Armor School.

On 1 May 1945, s.Pz.Abt 510 picked up two training Tigers in Putlos and were in action against British patrols, on 5 May 1945.

The British occupied the Gunnery School, on 8 May 1945, and later it was used as a RAF base.

To date I have found no German records as to the fate of these two Jagdtigers,[1] or any evidence from Allied forces that they remained in Putlos at the wars end.

40.3 Jagdtigers with the 17th Reserve Training Battalion Freistadt

Two Jagdtigers transferred to Freistadt training camp, on 8 March 1945.

A report stated:

Pz.Jg.Ers und Ausv. Abt. 17 Freistadt den 12 March 1945
From Gen Insp of the Pz troops Origin (K) Berlin

The Battalion Report

Day of the arrival	*8 March 1945*
2) Journey number	*6842681*
3) Number of vehicles (2)	*chassis No 305002 and 305003*

These two Jagdtigers were in Freistadt a matter of weeks before they were ordered to transport to Nibelungen Werk for rework and eventual deployment with 6 SS Pz Armee.

The 6th SS Panzer Armee was made up of the remnants of eight Panzer divisions including "Leibstandarte Adolf Hitler", "Das Reich", "Hohenstaufan" and "Hitler Jugend". After the abortive attack at Lake Balaton 6 - 15 March 1945, this force had been fighting a withdrawal back towards Vienna and St Polten.

An order,[2] in April 1945, placed the Jagdtigers at Freistadt under the command of 6th SS Panzer Armee. These Jagdtigers were to be returned to the factory for repair/action.

The Jagdtigers were put on rail transport and sent back towards Linz. In spite of all precautions, the train was seen and attacked by U.S. fighter-bombers just before Mauthausen, 5km east of Linz. The Jagdtigers were destroyed [3]. An eyewitness, Karlheinz Flick stated that **two** or **three** Jagdtigers [4] were on the train.

Oblt Karl Seitz who was formerly with the Stabs Kp s.Pz.Jg.Abt.653 until he became an instructor at Freistadt/Austria in autumn 1944, corroborated this, however in his account

Plate 349. Jagdtiger traveling east from Linz, in early May. It has no crane mounts – this was the Jagdtiger formerly damaged in the air raid, on 23 March 1945 (Wolfgang Schneider).

Plate 350. Jagdtiger "Sunny Boy" in camouflage position awaiting Soviet attackers. It has a motif painted on the front left superstructure plate, obscured by netting (Wolfgang Schneider).

dated 30/8/90, he stated **four** Jagdtigers[5] from Freistadt destroyed on the Train!

Only further documents or photographs of the destroyed train can prove the actual number of Jagdtigers involved, **is any one able to help?**

40.4 Work continues in Nibelungen Werk

As previously mentioned Heinz Grien from the s.Pz.Jg.Abt. 512 had been stationed at Nibelungen Werk, since the end of February 1945. He was to oversee Jagdtiger production and repairs; he reported directly to his Battalion commander, Major Walter Scherf.

Grien stated that all had gone well at the factory until the air raid, on 23rd March, when all the 150 metric ton cranes had fallen down and repairs and production could not proceed for a further 14 days.

In his last report that he wrote to Scherf, he stated that no ammunition was available for the four completed Jagdtigers and they would not be dispatched without ammunition.

He was threatened with court martial.

On 15 April, **four** Jagdtigers were recorded as complete.

On 29 April, these were transferred to a detachment from s.Pz.Jg.Abt.653; these were all seasoned veterans from the Battalion, which at the time was rapidly running out of Jagdtigers.

Grien recalls **nine** Jagdtigers at the factory at the end of the war. He stated that on 29 April, they were ordered by the inspector General of Panzer troops to deploy to St. Polten where they were put in the front line near Prinzersdorf [6]. On 8 May, the Americans captured them!

40.5 New equipment in April

On 15 April, a detachment from s.Pz.Jg.Abt.653 took on four new Jagdtigers from the factory near Linz. The troops had 2 weeks with these vehicles before, on 29 April, they were ordered to drive into action to counter the Soviet forces striking towards Linz.

Plate 351. The last three Jagdtigers photographed when the American and Soviet forces linked up in Strengberg, on 5 May 1945. The Jagdtiger in front is the same vehicle as in plate 349 (U.S. Army).

Plate 352. A fitting photograph to depict the end of the war and the combat operations of the most powerful armored vehicle to see service. An American soldier poses with his Soviet counterpart on top of the middle Jagdtiger in the line of three. Its motif, a toy bear is clearly visible (Bill Auerbach).

On 29 April, the last report was sent from Nibelungen Werk:

1. Tank deliveries since 15 April, 4 (128mm) Jagdtigers.
2. Expected deliveries by end of April are 4 (88mm) Jagdtigers.[7]
3. 17 Jagdtigers expected to be built by end of May, all 88mm.

A further detachment, including Hans Knippenberg, were making their way to Linz to take on the new Jagdtigers (88mm) expected to be ready, at the end of April 1945.

Seven/Eight Jagdtigers stood ready and were being guarded outside the factory. This new batch of Jagdtigers was fitted with 88mm gun [8] and the detachment were busy trying to secure ammunition. The guns had not been test-fired and consequently they had not been calibrated. There was still no optics available for the new version Jagdtigers.

The four 128mm Jagdtigers had not been put through gunnery trials and, on 30 April, they drove east from Linz towards St. Polten.

To-date, my attempts to obtain photographic proof of the 8.8cm Jagdtiger from German and Russian sources has proved fruitless! **Is any one able to help?**

40.6 The four Jagdtigers in combat

The **four** new Jagdtigers were ordered, on 29 April, to leave the Nibelungen Werk area and proceed due east towards St Polten, to counter advancing Soviets forces. These were released without the cannons having been test fired; there had not been the opportunity or time available to ship them out to Döllersheim for this purpose.

Four Jagdtiger crews from 653 were assigned to these new vehicles and formed a Kampfgruppe commanded by a Hauptsturmführer from the LAH.

The four Jagdtigers under heavy camouflage drove along the road from Linz through Enns, Strengberg, Amstetten, as far as Wieselburg, here they were loaded onto railway flat wagons.

The train only took them as far as St. Leonhard. On the morning of 1 May, there was a skirmish with Soviet tanks pushing south from St. Polten; the Jagdtigers repulsed this attack, however, they had realized that the Soviets had already broken through the front!

The Jagdtigers were driven back along the St. Polten Linz road, near Newmarket, one Jagdtiger "Sunny Boy" commanded by Fw Golinski broke down on the Ybbs River Bridge, and the crew abandoned it and jumped onto the other Jagdtigers.

In Amstetten the 3 Jagdtigers drove over a barricade in the market place, the Americans and Russians who were unable to stop the small Kampfgruppe already occupied the town. They continued to drive eastward at speed through Oed as far as a sharp bend in the road at Strengberg, there was a omnibus and two Russian tanks across the road.

The Jagdtigers stopped and surrendered to the Americans, the photo opportunity was taken.

One of the three Jagdtigers photographed at the back of the line was later taken to Kubinka for evaluation. (See Chapter 42/3) Fw R. Schlabs had commanded it.

The crews had painted cartoon motifs on the front left superstructure plates for vehicle identification; they had not used the usual three number system, as there were not enough Panzers to necessitate this!

40.7 The last two Jagdtigers in action?

There is a questionable account alleging[9] the acquisition and use of two Jagdtigers by the remnants of the schwere SS-Panzer Abteilung 501 in early May 1945, here is the thus far unsubstantiated account.

On 2 May, 40 troops from the schwere SS-Panzer Abteilung 501 were sent from Scheibs to the Nibelungen Werk. They had orders to make 6 Jagdtigers operational. Over the following 2/3 days, **two** Jagdtigers[10] were made ready for action.

On 5 May, both Jagdtigers intended to drive straight along the Linz - St. Pölten road. However, they were informed that this road was now in Allied and Soviet possession, they had to drive south east through Waidhofen, Gresten to Scheibs a total drive of about 50km, which took almost two full days. By 7 May, both Jagdtigers reached the divisional command post at Scheibs. They were deployed together with several Pz IV's, and were ordered to move off for action in the direction of Enns, back along the road they had just traveled! That night they reached Waidhofen a 25km road march.

On 8 May, the tanks were called back to cover the withdrawal from the Soviet troops, Scheibs had been over run by Soviet armor and all remaining resistance in the area was folding very quickly. One Jagdtiger broke a track when driving over a bridge crossing the Ybbs River,[11] it had to be pulled off the road. The other tanks took up positions at the edge of Waidhofen, from where they could control Weyer Markt for further withdrawal.

The advancing Soviet tanks were ambushed, burning Soviet tanks littered the road, 4 Pz IV's were lost in the vicious fire fight that followed, the attack was suspended for that day.

As darkness fell the small panzer group withdrew 12km to Weyer Markt, the following day 9 May, the remaining Jagdtiger was placed in the middle of the narrow street, it and the remaining Pz IV's were blown up to prevent the Soviet tanks from getting through the village. The troops disengage across the river Enns near Losenstein, and in the evening surrendered to the Americans near Steyr. That same day the Soviet forces occupied St. Valentin and the Nibelungen Werk.

40.8 New Jagdtigers not deployed

The Nibelungen Werk had continued to build Jagdtigers, until 3 May 1945. **Seven/eight** Jagdtigers had been completed and had been taken outside the factory where they were guarded to prevent sabotage. All were 8.8cm versions, a modification due to the lack of 12.8cm gun mounts arising from heavy bomb damage followed by the loss of the Krupp Berta Werk AG in Breslau. Work on the 8.8cm version Jagdtigers had started, in March 1945 and these vehicles were complete except for the gun sights, without which they were useless.

On 2 May, a detachment from s.Pz.Jg.Abt 653 under Lt. Hans Knippenberg was in Nibelungen Werk waiting for orders to take on these Jagdtigers. The order was never given.

On 4 May, with the Americans already near Linz, the order was given to blow up these new Jagdtigers to prevent capture. Lt. Hans Knippenberg's men, who had become quite experienced at blowing up Jagdtigers, carried out this order. Knippenberg states that **seven** or **eight** Jagdtigers were blown up. There was no mention of any Jagdtigers taken over for action by the schwere SS-Panzer Abteilung 501!

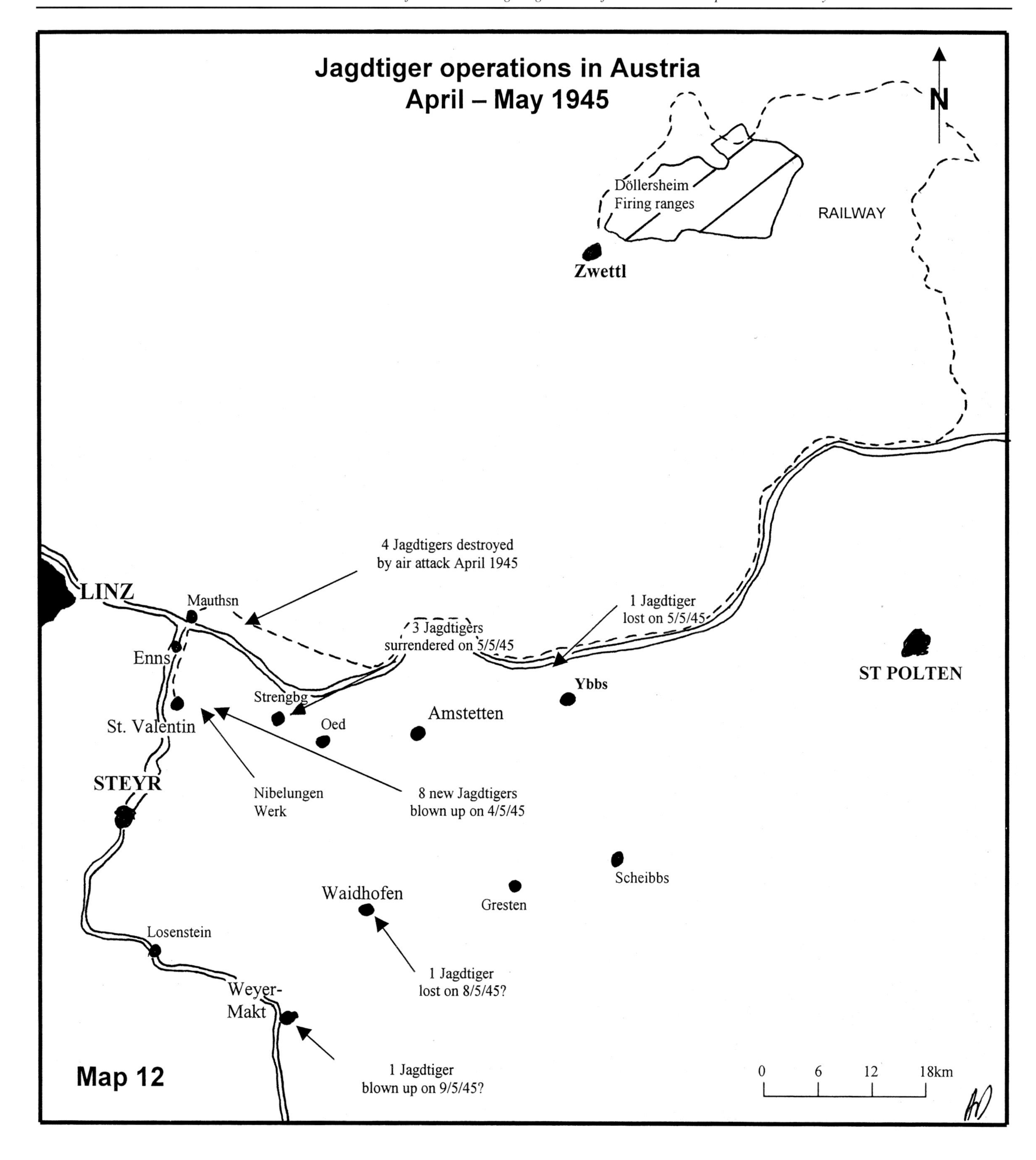
Jagdtiger operations in Austria
April – May 1945
N
Döllersheim
Firing ranges
Zwettl
RAILWAY
LINZ
Mauthsn
4 Jagdtigers destroyed
by air attack April 1945
3 Jagdtigers
surrendered on 5/5/45
1 Jagdtiger
lost on 5/5/45
Enns
St. Valentin
Strengbg
Oed
Amstetten
Ybbs
ST POLTEN
STEYR
Nibelungen
Werk
8 new Jagdtigers
blown up on 4/5/45
Waidhofen
Gresten
Scheibbs
Losenstein
1 Jagdtiger
lost on 8/5/45?
Weyer-
Makt
Map 12
1 Jagdtiger
blown up on 9/5/45?
0 6 12 18km

40.9 Jagdtiger deliveries in April 1945

Delivered to	Transported on	No	Type	Chassis No
s.Pz.Jg.Abt.653	*29/04/45*	*4*	*(H) 12.8cm*	3050??
			(H) 12.8cm	305082
			(H) 12.8cm	305083
			(H) 12.8cm	305084

40.10 Jagdtiger built in April 1945

There were 7 Jagdtigers completed against the program of 20 up to 30 April 1945.

3	*(H) 12.8cm*	Chassis No	305082
			305083 [12]
			305084
	4 (H) 8.8cm	Chassis No	305078
			305079
			305080
			305081

Note: 15 April 45, Ni-Werk stock 4 Jagdtigers, these completed in March/April

40.11 Hulls built by Eisen Werk in April 1945

There were 2 Hulls completed in April giving a final total of 116.

40.12 Jagdtigers built in early May

305085
305086 — If Completed
305087
305088

40.13 Conclusion

It is difficult to accurately establish the exact numbers of Jagdtigers and sequence of events, at this concluding period to the war!

The only theory that fits all three accounts, is that Heinz Grien did count nine Jagdtigers after the four 12.8cm Jagdtigers had already been dispatched, two were taken on by the schwere SS-Panzer Abteilung 501, this conveniently leaves the seven that Hans Knippenberg's men blew up. However this only satisfies the mathematical aspects of the accounts and not full historical accuracy.

I personally would add most credence to the R. Schlabs account, which is in part substantiated by photographs, as well as the release document. He only names the commander of a second Jagdtiger, therefore the other two could have been staffed by troops from the SS-Panzer Abteilung 501! The number of Jagdtigers available to have been blown up by H. Knippenberg is 50% substantiated by the release document and given time could possibly be fully substantiated by Russian intelligence photographs.

Photographs taken by the citizens of Waidhofen or Weyer Markt could prove the schwere SS-Panzer Abteilung 501 account in that area!

I would be delighted to correspond with any one who has access to further information and documents or photographs concerning this matter!

(1)The Two Jagdtigers are recorded as transferred Putlos and therefor this can be treated as a fact. It is likely that an attempt was made to return these vehicles to the Nibelungen Werk! **Can any one assist to verify this?**
(2)Order not found!
(3)Source Uffz KH Flick, Training Battalion Officer, 17th Reserve Training Battalion witness to the air attack.
(4)It is not known if these Jagdtigers or any parts were collected and sent the 8km to back to Nibelungen Werk, this would have depended on the severity of damage and the availability of recovery equipment, it is certain that the line would be cleared and repaired.
(5)If number correct, Jagdtiger 305001 and the other Porsche Jagdtiger removed from the 653 inventory will have been on this train!
(6)This location virtually ties in with R. Schlabs account, he does not state how many Jagdtigers actually left the factory!
(7)This was the last report sent from Nibelungen Werk and to-date is the only piece of documented evidence as to the existence of these vehicles!
(8)Speculation, records can prove only four!
(9)Only documents or photographs tied to these locations mentioned could confirm this.
(10)If true it is highly likely that these were 12.8 cm Jagdtigers, they could have been vehicles returned from Putlos or the missing 305001 returned from Waffen Pruef 6 and the other Porsche Jagdtiger formerly with s.Pz.Jg.Abt. 653 in Döllersheim, or even the 2 which did not get through to 3/512 in Sennelager?
(11)This statement is far too coincidental with the loss of Fw Golinski's Jagdtiger in R. Schlabs account! **Make your own judgment!**
(12)This now in Kubinka.

41

After the Battles

41.1 Allied intelligence on the Jagdtiger

British Military Intelligence were aware of the Jagdtigers existence almost as soon as the wooden mock up was shown to Hitler at the Arys Proving Ground, in East Prussia on 20 October 1943. A high priority to determine the technical design of the Jagdtiger was assigned as of the Allied technical intelligence, in the last year of World War II.

In late 1944, MI 10 obtained the answer to this question, in the War Office, from a captured notebook, containing excellent pencil drawings of most of the then current German tanks and self-propelled guns. The Jagdtiger illustration had as usual been torn out, but some of the pencil had rubbed off on the back of the preceding page. Scotland Yard's Forensic Laboratory was able to bring out the mirror image of the crucial drawings. This information was quickly circulated to all Allied operational units.(1)

14 February 1945:
Notes on Tiger Jäger (Jagdtiger) and on tank production at Linz Eisen Werk.(2) This followed the interrogation of a prisoner who had been a foreman welder at the factory.

2 March 1945:
US 12th Army group report on first captured Jagdtiger.(3)

10 May 1945:
Report Jagdtiger Intelligence by Lt.Col. GC Reeves and Mr. Goldthorpe on Jagdtiger, 2 vehicles observed and photographed at Eppingen.(4)

10 May 1945:
Report No.275 Observations by Lt. W B Curtis, Jagdtiger. Vehicle photographed at Netphen.(5)

Plate 353. Side view of X1, the Jagdtiger formerly commanded by Albert Ernst (U.S. Army).

Plate 354. Rear view of Jagdtiger X2 after it had been driven to Menden Camp, on 17 April 1945 (U.S. Army).

Plate 356. Jagdtiger on a parade ground in Iserlohn. It is not known which of the three it was (Karlheinz Münch).

After the War, actual evaluation on captured vehicles was carried out in Sennelager, the vehicles then being sent to England, the Americans used Aberdeen Proving Ground and the Soviets Kubinka in Russia. All 3 vehicles evaluated have survived to this day.

An Allied inspection team was eventually given permission to visit the Nibelungen Werk in July 1945 however by this time all the records had been removed from the factory!

41.2 The 1/ s.Pz.Jg.Abt.512 Jagdtigers after capture

After the 16 April, Jagdtiger Unit 512 no longer existed, all had gone into captivity.

The 3 surviving Jagdtigers of 1/512 (Kp Ernst) were driven for their captors from Iserlohn to the U.S. Army Camp near Menden, where the German troops were held in captivity. Here they were displayed for the U.S. troops who had not seen anything so powerful. Again they were photographed.

One of these Jagdtigers [6] was taken to an Allied military base in Iserlohn and stored at a troop parade ground. It was photographed with a Belgium soldier after the war.

I do not know what happened to these Jagdtigers after this, however, many captured German tanks were transported back to Sennelager Camp which was taken over as a base by the British Army, in 1945.

The ballistic properties of the tanks were often tested when they were used on the ranges as hard targets for the British tank gunners.

As can be seen from the photos, the Jagdtigers generated a lot of curious interest from the American troops.

One Jagdtiger did survive at Sennelager for over 10 years after the end of World War II.

41.3 Wreck removal after the War

The destruction prevailing in Europe after World War 2 was appalling, war debris, ruined buildings, unexploded ordnance and displaced peoples, being the major issues for the victorious occupying powers. These had to be tackled in the immediate years following the end of hostilities.

Allied military personnel were used, supplemented by the local civilian population, in the clear-up and rebuilding program. It was a massive task.

Plate 355. Side view of the same vehicle (U.S. Army).

Plate 357. This attempt to remove Jagdtiger No 131 failed (Karlheinz Münch).

Literally thousands of wrecked military vehicles, many with dangerous ammunition, had to be cleared to allow life to return to something approaching normality.

The heavier wrecks were the most difficult to deal with. Often many recovery vehicles were used, to drag them to assembly areas. An inspection was made before the scrap mans cutting torches quickly reduced them into manageable pieces for recycling.

It took almost 5 years to clear the battlefields, in Europe.

Every Jagdtiger in Europe, except the 3 removed for evaluation was scrapped, the last one in Sennelager, in the late 1950s.

[1]Source Peter Gudgin.
[2]See Chapter 1, Drawings 2 - 6.
[3]See Chapter 31, Plates 238 - 244.
[4]See Chapter 34, Plates 290, 293 - 297, 299-301.

Plate 358. Photo taken in Eppingen in June 1945. Jagdtiger tactical number 123 has already started to be cut up on site. Its fighting compartment has been removed (Karlheinz Münch).

[5]See Chapter 38, Plates 313 - 318.
[6]The tactical x-number as yet is not known on this Jagdtiger?

Plate 359. Front view of Jagdtiger in Sennelager; it belonged to 3/512 and has the crane mounting brackets. This is thought to be the same vehicle as in plates 321-326 (Tank Museum).

Plate 360. Side view of the same vehicle, photos taken in 1955 (Tank Museum).

Plate 361. Rear view of the Jagdtiger. Unfortunately it did not survive the cutting torch (Tank Museum).

42

The Surviving Jagdtigers

42.1 Jagdtiger (305004) in England

This Jagdtiger built in July 1944 was released to Wa-Pruef 6 in August 1944, as the replacement for Jagdtiger 305001, which had suffered major mechanical problems during testing. The Nibelungen Werk release record was corrected on 1 September 1944 to record this. Throughout 1944 and early 1945, it had been used as a test vehicle (No 253) by Wa-Pruef 6, in Haustenbeck-Sennelager.

It is recorded in the inventory as one of the Panzers of Panzer Gruppe Kummersdorf on 31 March 1945 (Chapter 38.15). It is thought to have seen action in Sennelager and was first photographed by the Allies on 6 April 1945, by a photographer from the Keystone Press Agency.

After its capture it was taken over by the British, in 1945. It was first evaluated in Kummersdorf in mid 1945, when it was still driveable. These tests were filmed, and part of this sequence can be seen on the video "Axis Armour" produced by Command Vision Ltd., PO Box 393, 7-11 Britannia Place, ST. Helier, Jersey, Channel Islands.

It was later transported to England, on a Gotha trailer, and was tested at the Chertsey Camp for intelligence on captured armor. Unfortunately, its central engine deck-plate remained at Sennelager along with the broken suspension unit.

Plate 362. Right side view of 305004 in Sennelager (Tank Museum).

Plate 363. Front view of 305004 (Tank Museum).

Later, it was moved, in the late 1950s, to Bovington Camp Dorset, where it resides today as one of the largest exhibits in the Tank Museum.

The vehicle is in excellent condition and is complete except for a suspension unit which broke-off before capture, its engine deck cover-plate is also missing. It would not need much work to get it running. The chances of finding replacement suspension units other than by a costly remanufacture are to say the least remote!

The vehicle was originally released from the Nibelungen Werk painted in basic dark yellow and remained that color until its capture. Later, at Bovington, it was painted dark green, then in the 1960s some of its roof fittings were cut off and handrail brackets were fitted to allow visitors to climb on to it. The handrail has since been removed but not the mounting brackets.

Plate 364. Left side view clearly shows the incomplete suspension system (Tank Museum).

Plate 365. British troops put 305004 through its paces in Sennelager (Tank Museum).

Plate 366. Jagdtiger 305004 arrives at Bovington camp (Tank Museum).

Plate 367. Photographed in 1972 the hand railings are clearly being used (Author).

Plate 368. Author and Jagdtiger in 1982 (Author).

Plate 369. Author's brother Ian, who helped to measure the vehicle and assisted in sawing the steel plates for the model (Author).

Plate 370. The remains of the Zimmerit are still evident (Author).

Plate 371. Close up of engine (Author).

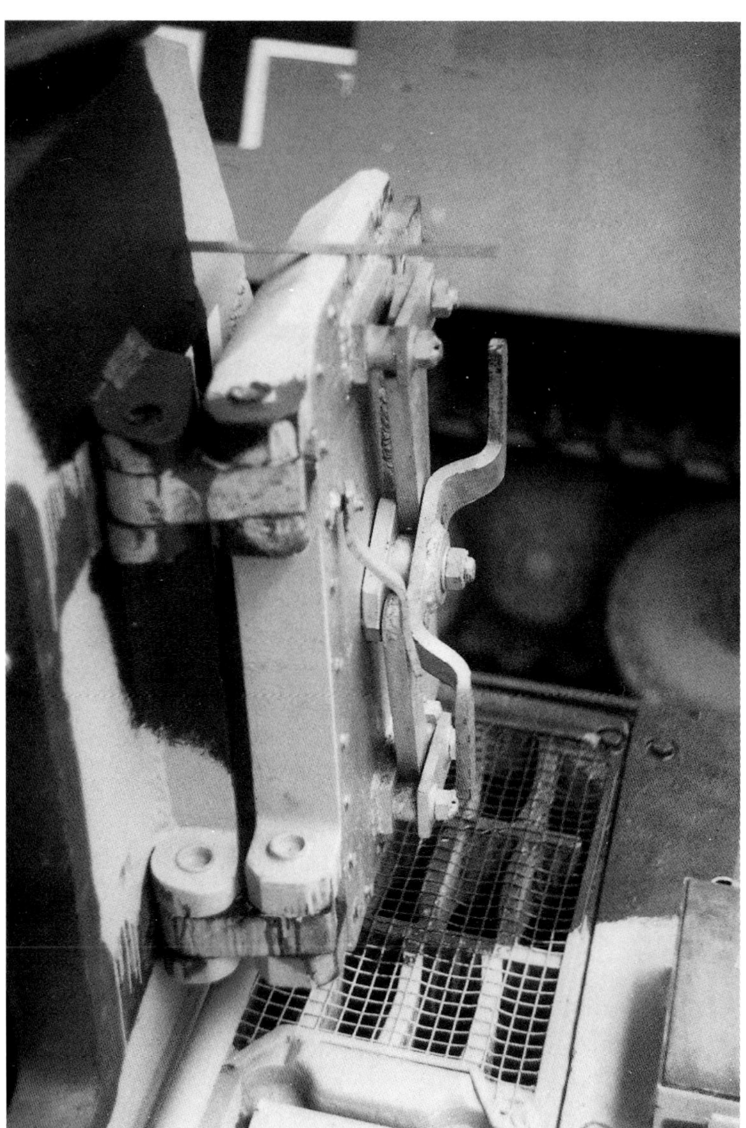

Plate 373-374. Close ups of the right side rear door with 12 inch steel rule (Author).

Plate 372. Close up of commanders hatch (Author).

Plate 379. Side stowage hooks and fittings (Author).

Plate 375-376. Rear 3/4 views of 305004 (Author).

Plate 377-378. Saukopf (Author).

Plate 380. Front view (Author).

Plate 381. The exhaust cover plating still remains (Author)

Plates 382-383. Close-ups of the ground plate for the headlamp (Author).

It now supports a very clean inauthentic three-color camouflage scheme but is in a relatively good condition.

42.2 Jagdtiger (305020) in America

This Jagdtiger built in October 1944 was released as one of three from the Nibelungen Werk on 24 November 1944, and it arrived in Dollersheim on 28 November 1944. It was assigned to 3/schwere Panzer Jäger Abteilung 653 and given the tactical number 331.

After the furious battle in Neustadt an der Weinstrasse, between elements of the U.S. 10th Armored Division and a Kampfgruppe from 3/schwere Panzer Jäger Abteilung 653. Two of the three Jagdtigers were captured, both had received multiple shell hits on their frontal armor.

Jagdtiger No 331 had taken six hits on its front end, it was captured along with Jagdtiger No 323 which had taken four hits. No 331's main armament had been put out of action when the crew part drained the recoil cylinder before firing the last round. This wedged the gun back into its full recoil position, immediately afterwards, it was abandoned by its commander Lt. Casper Geoggerle and his crew.

The Jagdtiger was factory painted in an unusual camouflage scheme, the colors being sand, red-brown, olive green and field gray. It had the numbers 331 painted on each side of its fighting compartment. Still visible was its chassis number, 305020, which was stenciled in black numbers on its nose plate.

This vehicle was recovered by the Americans and shipped to the U.S. for evaluation at Aberdeen Proving Ground, where it still stands today. The vehicle is still complete with the exception of the front sprocket teeth rings, which have been removed. Unfortunately, its current camouflage color, sand and gloss brown are definitely not authentic, which spoils the look of the vehicle. Since it has stood outside for over 50 years, internally and mechanically it is in a poor state, but certainly not beyond a major restoration project!

The French dragged the second Jagdtiger No 323 to the Festival Platz in Neustadt where it remained until it was cut up for scrap in 1948.

Plate 384. Jagdtiger 305020 arrives at Aberdeen Proving Ground on rail transport without tracks (Aberdeen Proving Ground).

Plate 385. It is photographed in its first display area (Aberdeen Proving Ground).

Plate 386. Photographed in the late 1940s, its chassis number 305020 was painted in black just above the white FMAR 1017 which was a U.S. intelligence number (Aberdeen Proving Ground).

Plate 387. A good comparison of the Jagdtiger with an Elephant (Aberdeen Proving Ground).

Plate 388. In the 1970s it received an unusual speckled paint coat (Aberdeen Proving Ground).

Plate 389- 392. Photographs of 305020 in overall dark yellow color scheme. It has a right side skirt plate over its exhaust pipes (Aberdeen Proving Ground).

Plate 393 - 397. In its most recent color scheme, 50 years on, it still bears the battle scars from protecting Kasper Geoggler and his crew (Per Sonnervik).

Plate 398-399. Internally the vehicle is in a poor condition (Aberdeen Proving Ground).

42.3 Jagdtiger (305083) in Russia

Built in April 1945, this was one of four 12.8cm Jagdtigers recorded as being complete on 15 April 1945. These were subsequently released for combat on 29 April 1945.

After a very short period of action under the operation of Fw. Reinhold Schlabs and his crew, it was one of three surrendered to the link up between the Soviet and Allied Armies which occurred, on 5 May 1945, in Strengberg. All three Jagdtigers were operational.

Jagdtiger No 305083 was then taken to the Kubinka School of Armor and Evaluation, the Russian center for armor intelligence, situated 15 miles south of Moscow. The preserved vehicle was only part covered for many years before being placed in Hanger No 6, where it stands today between a Krupp Ardelt SP 88mm and a Tiger II (Series Turret).

The Jagdtiger has its full skirt armor and has been recently resprayed. All stowage has been removed and a new Russian headlamp has been fitted. The vehicle is not a runner but it is complete except for an air filter unit and its optics. The hatches have been left open. Written in welding on its front glacis plate it actually states that it was captured, on 5 May 1945.

It would require considerably less work than 305020 to get it running.

A careful study of the three photographs showing the Jagdtiger being tested at the Kubinka Proving Grounds reveals some interesting features:

1. A detailed study of the position of the sprayed three-color camouflage scheme proves that this is not Sunny Boy, Teddy Bear, or the Jagdtiger at the front of the surrender column. This analysis reveals that it was the Jagdtiger at the back of the line of three and commanded by Fw Reinhold Schlabs.

2. The vehicle is fitted with the monopod mounting plate for the M.G. 42 on its engine deck.

3. There is an unusual bracket fastened to the back plate, between the exhausts which was for recovery purposes, this was later broken by the Russians during testing.

4. There is no visible evidence of any battle damage, even the skirt plates are not bent.

Plate 400. Front view of 305083 while under test in Kubinka Proving Ground (S. Netrebenko via Thomas Anderson).

Plate 401. From the left side it looks nose heavy. It was painted in the three color camouflage scheme (S. Netrebenko via Thomas Anderson).

Plate 402. Rear view of 305083. It was fitted with an unusual anchor bracket for a central pull – this later broke during evaluation (S. Netrebenko via Thomas Anderson).

Plate 403. The central towing bracket is now broken. (Author).

Plate 406. Close up of the welding holding the angle steel for the roof plate (Author).

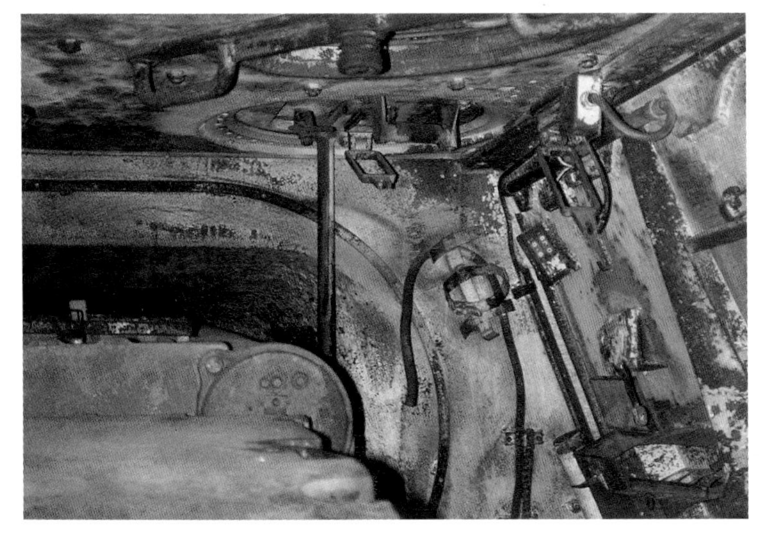

Plate 404 - 405. Internal photos either side of the gun (Author).

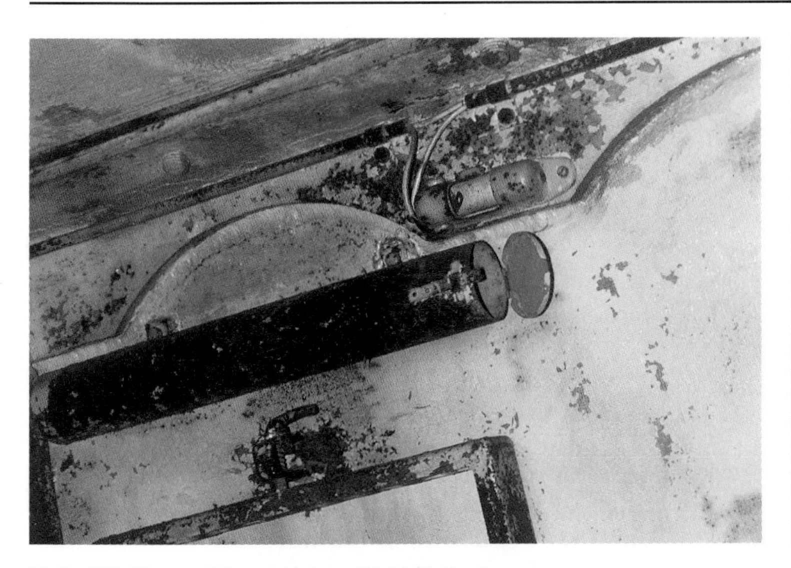

Plate 407. Flag canister and internal light (Author).

Plate 408. The Draeger ventilator has been removed (Author).

Plate 409. The screw clamp for the internal gun lock (Author).

Plate 410-411. Fuel and water filler caps (Author).

Plate 412-413. Engine deck details (Author).

Plate 414-415. Central engine hatch details (Author).

Plate 416-417. Close up views of the very late type front mud guards (Author).

Plate 418. Written in weld and highlighted in white paint it states that it was captured on 5.5.45 (Author).

Plate 419. Right side view (Author).

Plate 420. Stowage for projectiles (Author).

Plate 421. Internal view of drivers position (Author).

Plate 422- 424. Comparison of the three Jagdtigers preserved to-date (Author).

43

Author's 1:12 Scale Model

My interest in German armor, which had been developed through several years of plastic modeling, progressed in 1980, into a model-engineering quest. The project I chose was to build a 1: 12 scale version of my favorite Jagd-Panzer "JAGDTIGER".

This would require using scale thickness steels and other metals for the construction, the model was planned to be driveable and also to be able to fire, permitted on a Section One License! Little did I know that to achieve this, the work would take some 8000 hours, spread over an eleven-year period! The research toward the model, progressed into a much greater passion, a hunger for knowledge from both the technical and tactical aspects, about the Jagdtigers themselves!

As mentioned in the introduction, in 1980 I embarked upon a model-engineering project to build a 1:12 scale-working model of a Jagdtiger. With this undertaking I was able to acquire the necessary machining skills to produce an award-winning model.

MODEL SPECIFICATIONS

Length	0.9m.
Width	0.3m.
Height	0.25m.
Weight	75kg.

MODEL FEATURES

It is built out of scale-thickness steel plates, welded, with interlocking joints of correct proportions. The model also features independent suspension, track-tension, drive (electric) from front-sprockets, and braking.

The gun has hand-traverse and elevation through a correct angle of sweep; since 1985 it has also been licensed to fire 0.410" shotgun cartridges, hence it has full recoil and recuperation facili-

Plate 425. Author's model in competition at the 1992 Model Engineers Exhibition (Author).

Plate 426. Author and trophies in 1992 (Author).

Plate 427. Right side view with spare tracks removed (Author).

Plate 428. Rear view (Author).

Plate 429-430. Front and rear views (Author).

Plate 431. The detail of the steel tracks can be seen in this view (Author).

Plate 432. Top view (Author).

Plate 433-434. Photographs of Author's 1: 12 model in camouflage positions (Author).

Plate 435. Author in Jagdtiger 305083 taken in Kubinka in August 1993 (Author).

ties. The model also has full external details, and is fitted with steel-tracks pinned correctly right through their full-width.

In 1991, with this model, I won The Midlands Model Engineering Class 15, and in 1992 the Model Engineers Exhibition Gold Medal in Class E5, together with the Exide and Drydex Cup for the best battery-powered model of the show. **I will not make another one!**

I am prepared to correspond with any one who is considering or has started any similar armor-modeling project.

Bibliography

Primary sources:

1. Intelligence Reports (held at the Tank Museum, Bovington).

(a) *Technical Intelligence Report No. 275 (Jagdtiger),* 10 May 1945.

(b) *Notes on 'Tiger-Jäger' and on tank production at Linz Eisen Werk,* February 1945.

(c) *Intelligence Summary on Tiger II, 1945.*

(d) *Report on examination of German PzKpfw VI clutch*, by Vauxhall Motors Ltd.

(e) *Laboratory Report on shock absorber*, Morris Motors Ltd.

(f) *Intelligence Summary, gearbox, steering, final drives (Tiger),* November 1943

(g) *Tiger Report Part IV,*	Power Plant	September 1944
	Dynamo	April 1944
	Inertia-starter	March 1944
Part IX, Automatic fire extinguisher	January 1944	
	Ventilation	September 1944
Description of Henschel bogie wheel		June 1944.

Report on Porsche torsion-bar design standards 1945.

(j) *Report on German tank wireless systems.*

2. Werkstatt Handbuch zum Maybach – (motor HL 230 P30 - HL 210 P30).

3. Handbuch fur den Tiger II.

4. From the Bundesarchiv Military Archiv Freiburg: *(FILES OF THE GEN INSP D PZ. TR.)*

Reports from file RH 10/91	
Reports from file RH 10/118	pages: 22
Reports from file RH 10/121	pages: 1, 4 - 9
Reports from file RH 10/122	pages: 40
Reports from file RH 10/123	pages: 66
Reports from file RH 10/125	pages: 113 - 122, 164 - 165, 198, 199, 266, 315, 348
Reports from file RH 10/147	pages: 62 - 80
Reports from file RH 10/364	Pages: 68, 69

5. Service entry book for Jagdtiger 305023.

6. Correspondence with

a. Major Walter Scherf.

b. Oblt Otto Carius.

7. Via Wolfgang Schneider

a. Major Rolf Fromme.

b. Hptm Albert Ernst.

c. Lt Erich Schröder.

8. Via Karlheinz Münch

a. Lt Hans Knippenberg.

b. Fw Reinhold Schlabs.

c. Johan Schleiss.

d. Uffz Appel.

e. Uffz KH Flick.

f. Oblt Karl Seitz.

g. Obfw Heinz Grien.

h. Oblt Hans Fink.

Secondary sources: Published works:

AUERBACH, William. *Last of the Panzers,* Arms and Armour Press, London, 1984.

AGTE, Patrick. *Michael Wittmann und die Tiger der Leibstandarte SS Adolf Hitler,* Deutsche Verlagsgesellschaft, Rosenheim, 1994.

CARIUS, Otto. *Tigers in the mud,* J. J. Fedorowicz Publications, Winnipeg, 1992.

CHAMBERLAIN, Peter, DOYLE, Hilary L, JENTZ, Thomas L. *Encyclopaedia of German Tanks of World War Two,* Arms and Armour Press, London, 1978.

DOYLE, Hilary L, ELLIS, Chris. *Panzerkampfwagen German Combat Tanks 1933 - 1945,* Bellona Publications, Kings Langley, 1976.

DOYLE, Hilary L, JENTZ, Thomas L, SARSON, Peter. *Kingtiger Heavy Tank 1942-1945,* Osprey Publishing, London, 1993.

DOYLE, Hilary L, JENTZ, Thomas L, SPIELBERGER, Walter J. *Schwere Jagdpanzer,* Motor Buch Verlag, Stuttgart, 1993.

FLETCHER, David. *Tiger, the Tiger Tank, A British View.* H.M.S.O. London, 1986.

GUDGIN, Peter. *The Tiger Tanks,* Arms and Armour Press, London, 1991.

HOFFSCHMIDT, E.J., TANTUM, W. H. *Second World War Combat Weapons, German.* WE, Inc. Old Greenwich, 1968.

HOGG, Ian. *German artillery of World War Two,* Arms and Armour Press, London, 1975.

JENTZ, Thomas L, DOYLE, Hilary L. *Germany's Tiger Tanks,* Schiffer Publishing, Atglen, 1997.

KLEINE, Egon, KÜHN, Volkmar. *Tiger: Die Geschichte einer Legendären Waffen 1942 - 45.* Motor Buch Verlag, Stuttgart, 1981.

KNITTEL, Hartmut H. *Panzerfertigung im Zweiten Weltkreig,* Verlag E. S. Mittler & Sohn, Bonn, 1988.

KUROWSKI, Franz. *Panzer Aces,* J. J. Fedorowicz Publications, Winnipeg, 1992.

MAUS, Heinz. *"Panzerkampfwagen VI, Königstiger" - Dokumentation -,* Unpublished, Munster, 1989.

MUES, Willi. *Der große Kessel,* 1986.

MÜNCH, Karlheinz. *Combat History of the Schwere Panzerjäger Abteilung 653,* J.J. Fedorowicz Publications, Winnipeg, 1998.

PALLUD, Jean Paul. *Battle of the Bulge; Then and Now,* After the Battle Publications, London, 1984.

PAWLAS, Karl R. *Waffen Revue No 17,* Journal-Verlag Schwend, Schwäbisch Hall, 1975.

PAWLAS, Karl R. *Waffen Revue No 50 and 51,* Journal-Verlag Schwend, Schwäbisch Hall, 1983.

PERNY, P, SIEDEL, C, VOLTZENLOGEL, J, DEBS, A, WALTHER, C, POMMOIS, L.M. *Operation Nordwind,* Editions Heimdal, 1992.

SPIELBERGER, Walter J. *Panzerkampfwagen Tiger und seine abarten,* Motor Buch Verlag, Stuttgart, 1977.

SPIELBERGER, Walter J. *Panzerkampfwagen Panther und seine abarten,* Motor Buch Verlag, Stuttgart, 1978.

SCHNEIDER, Wolfgang. *Elephant - Jagdtiger Sturmtiger (Band 99),* Podzun Pallas Verlag, Friedberg, 1986.

SCHNEIDER, Wolfgang. *Tigers in Combat,* J.J. Fedorowicz Publishing, Winnipeg, 1994.

TANK MAGAZINE. *WWII Military Vehicle Photo File Vol 1,* Delta, Tokyo 1993.

WHITING, Charles. *Battle of the Ruhr pocket,* Pan/Ballantine, London, 1970.

WISE, Terence. *D -Day to Berlin,* Arms and Armour Press, London, 1979.

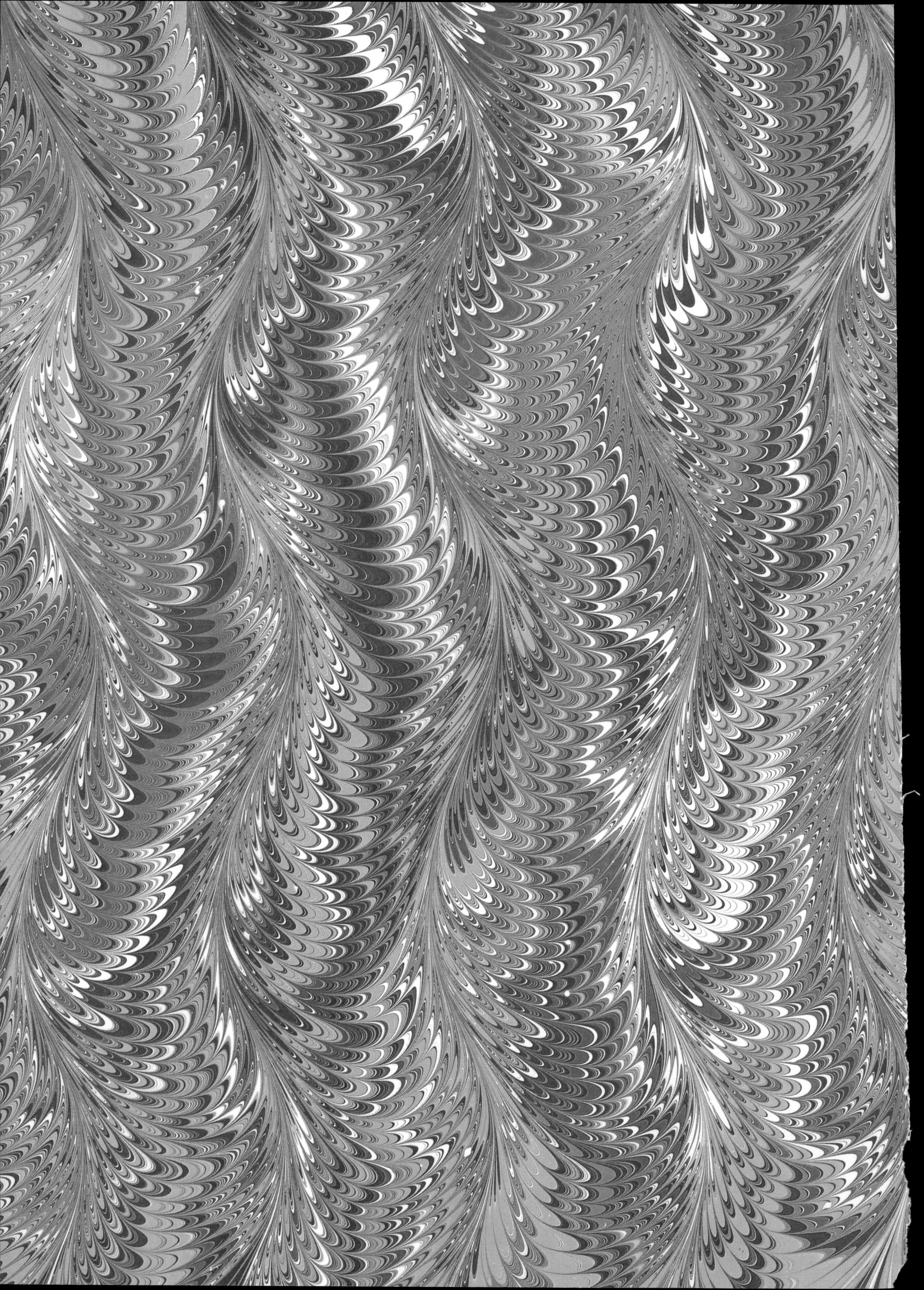